Laboratory Manual

CHARLES H. CORWIN

American River College

FIFTH EDITION

Introductory Chemistry

CONCEPTS & CONNECTIONS

PEARSON

Prentice Hall

Upper Saddle River, NJ 07458

Editor-in-Chief, Science: Nicole Folchetti
Senior Editor: Kent Porter Hamann
Assistant Editor: Carol G. DuPont
Marketing Manager: Liz Averback
Assistant Managing Editor, Science: Gina M. Cheselka
Project Manager, Science: Maureen Pancza
Supplement Cover Manager: Paul Gourhan
Supplement Cover Designer: Victoria Colotta
Operations Specialist: Amanda A. Smith
Senior Operations Supervisor: Alan Fischer

© 2009 Pearson Education, Inc.
Pearson Prentice Hall
Pearson Education, Inc.
Upper Saddle River, NJ 07458

Printed in the United States of America

10 9 8 7 6 5 4

ISBN-13: 978-0-13-604301-0

ISBN-10:   0-13-604301-1

Pearson Education Ltd., *London*
Pearson Education Australia Pty. Ltd., *Sydney*
Pearson Education Singapore, Pte. Ltd.
Pearson Education North Asia Ltd., *Hong Kong*
Pearson Education Canada, Inc., *Toronto*
Pearson Educación de Mexico, S.A. de C.V.
Pearson Education—Japan, *Tokyo*
Pearson Education Malaysia, Pte. Ltd.

Laboratory Manual

FIFTH EDITION

Introductory
Chemistry

CONCEPTS & CONNECTIONS

Contents

EXPERIMENT

EXPERIMENTS

* Assigned Student Unknown

APPENDICES

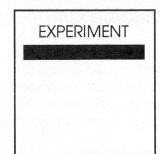

Preface

At a recent POGIL conference, an instructor using the *Prentice Hall Laboratory Manual* mentioned to me that the experiments were remarkably "bullet-proof." I replied that my colleagues and I instruct over 1000 students in the laboratory each year and our introductory chemistry program employs rotating adjunct faculty who bring a fresh set of eyes to the experiments we conduct each semester. This constant turnover presents a challenge to a smooth operation, but affords ongoing feedback and the opportunity to further fine tune each procedure and assignment. Moreover, the *Annotated Instructor's Manual* contains complete directions for stockroom personnel and provides a list of unknowns that yield good results.

The *Prentice Hall Laboratory Manual for Introductory Chemistry*, *5/e*, continues to evolve as pedagogical, environmental, and safety considerations advance. In this edition, particular attention has been paid to the environmental issue. This version does not contain any procedures that involve lead, mercury, chromium, chloroform, or carbon tetrachloride. Moreover, the experiments utilize 13 × 100 mm test tubes to further reduce chemical waste. No special equipment is required, and the experiments are *not* microscale.

The successful format found in the author's previous laboratory manuals has been retained, and the highly effective prelaboratory and postlaboratory assignments have been updated. The format for each experiment includes

- A set of **Objectives** to help students focus on the laboratory activities.
- A **Discussion** of the experiment, including example exercises to help students with calculations, and/or chemical equations.
- A list of **Equipment and Chemicals** to organize the necessary materials.
- A stepwise **Procedure** to guide student activities in a systematic fashion.
- A **Prelaboratory Assignment** to prepare students before coming to laboratory, and to alert them to any safety precautions.
- A **Data Table** to record observations and measurements.
- A **Postlaboratory Assignment** to reinforce the laboratory activity.

In addition, instructions for basic lab techniques are found in the appendices for the use of a laboratory burner, platform balance, beam balance, electronic balance, and a volumetric pipet. The appendices include **Answers to the Prelaboratory Assignments** to ensure that students are prepared before starting a laboratory experiment.

Laboratory Safety

Every effort has been made to ensure the safety of students in lab. Procedures that involve even minimal danger have been avoided. General **Safety Precautions** are listed in the front of this manual, and Experiment 1 has a safety quiz in the postlaboratory assignment. Students are alerted throughout to procedures and chemicals that should be performed carefully, and the prelaboratory assignments have questions regarding safety, which are noted in **Appendix J**.

Time Frames

Each experiment is designed to provide three hours of lab work, assuming ideal conditions. Since laboratories may differ—for example, the number of available balances—the time required may vary. In order to provide flexibility and reduce the time required to perform an experiment, the instructor has the option of eliminating a procedure or reducing the number of experimental trials.

Instructor's Manual and Quiz Item File

A complementary Instructor's Manual is provided with each adoption of the laboratory manual. The *Annotated Instructor's Manual* contains the following for each experiment: suggested unknowns and directions for dispensing and preparing solutions, sample data tables, answers to postlaboratory assignments, and a *Quiz Item File* containing over 500 class-tested questions.

The *Annotated Instructor's Manual* also contains a *Master List of Reagents & Suppliers* for all chemicals required for each experiment, along with directions for the preparation of aqueous solutions. A list of websites, addresses, and phone numbers of suppliers for chemicals and equipment is provided to assist stockroom personnel.

Acknowledgments

This latest set of experiments reflects the many suggestions of instructors and students who have offered comments. In addition, I am fortunate to have the shared expertise of colleagues including: Norm Allen, Constantino Aznar, Kristin Casale, Darren Gottke, Ron Grider, Robert Holmes, Tami Hong, Greg Jorgensen, Michael Maddox, Dianne Meador, Afarin Moezzi, Ed Niedzinski, Luther Nolen, Dustin Nouri, Karen Pesis, Forest Quinlan, Deboleena Roy, Daniel Stewart, John Terschak, Brian Weissbart, Veronica Wheaton, and Linda Zarzana. It has been a privilege to work with Carol Dupont, Assistant Editor at Prentice Hall, who has been invaluable in guiding the project through its many phases.

A successful laboratory program is helped immeasurably by competent stockroom personnel. The continual refinement of these experiments has been aided considerably by our stockroom lab technicians, Cuong Bui, Chris Douglas, and Ed Hege. I, along with my colleagues, greatly appreciate the support provided by Cuong, Chris, Ed, and staff.

I again welcome comments from instructors and students who perform these experiments, and I invite you to write to the address below, or e-mail me at **chcauthor@aol.com**.

Charles H. Corwin
Department of Chemistry
American River College
Sacramento, CA 95841

Laboratory Manual

FIFTH EDITION
Introductory Chemistry
CONCEPTS & CONNECTIONS

Safety Precautions

With the proper precautions, the chemistry laboratory is not a dangerous place. If you study the prelaboratory assignment before coming to lab, the laboratory should be no more dangerous than any other classroom. Most of the precautions are just common sense.

1. Wear safety glasses or goggles at all times while working in the laboratory.
2. Wear shoes at all times.
3. Eating and drinking in the laboratory are prohibited.
4. Locate the first-aid equipment and fire extinguisher.
5. Consider all chemicals to be hazardous unless instructed otherwise.
6. If chemicals come into contact with your skin or eyes, wash immediately with water and consult your laboratory instructor.
7. Never smell any vapor or gas directly. Instead waft a small sample toward your nose.

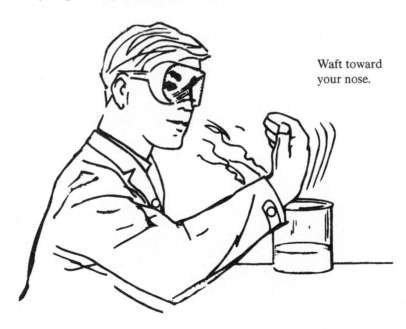

Waft toward your nose.

8. Any reactions involving noxious chemicals or unpleasant odors must be performed under a fume hood.

9. While heating a substance in a test tube, never point the open end toward yourself or your neighbor—the contents may erupt and spray a hot chemical.

10. Always pour acids into water—not water into acid—because the heat released will cause the acid to spatter.
11. Clean up all broken glassware immediately.
12. Many common organic reagents are highly flammable; for example, acetone, alcohol, and ether. *Do not use organic liquids near an open laboratory burner flame.*
13. Do not perform unauthorized experiments.
14. **Notify the instructor immediately in case of an accident.**

Locker Inventory

EQUIPMENT	QUANTITY
beakers, 100, 150, 250, 400, 600, 1000 mL	1 each
clay triangle	1
crucible and cover	1
crucible tongs	1
dropper pipet	1
Erlenmeyer flasks, 125 mL	3
Erlenmeyer flasks, 250 mL	3
evaporating dish	1
Florence flask, 1000 mL	1
long-stem funnel	1
small, plastic buret funnel (optional)	1
graduated cylinder, 100 mL	1
litmus paper, red and blue	1 each
stirring rod, thin glass	1
stirring rod with rubber policeman	1
test tubes, 16 × 150 mm	2
test tubes, 13 × 100 mm	6
test tube brush	1
test tube holder	1
test tube rack	1
thermometer, 110°C	1
wash bottle, plastic	1
watchglass, ~150 mm	1
wire gauze	1

ASSIGNED LOCKER # _____   NAME _____

COMBINATION _____   SECTION _____

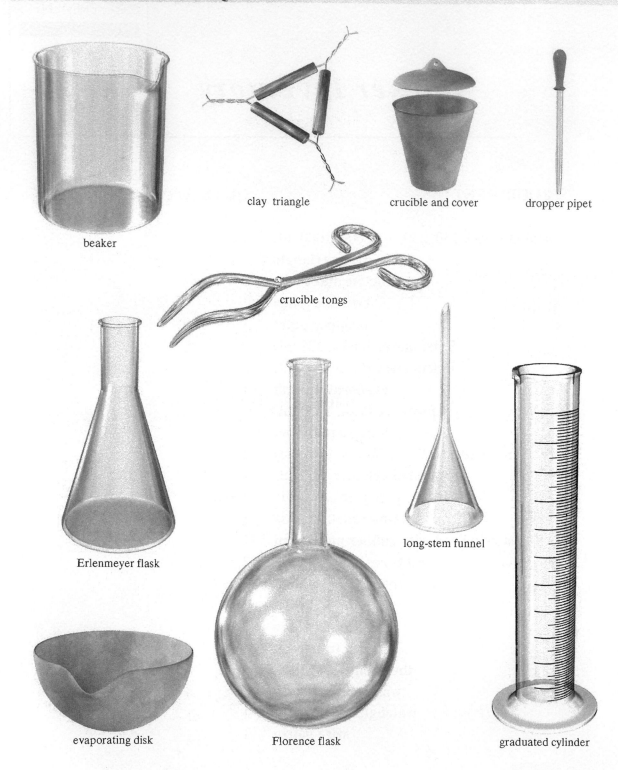

beaker

clay triangle

crucible and cover

dropper pipet

crucible tongs

Erlenmeyer flask

long-stem funnel

evaporating disk

Florence flask

graduated cylinder

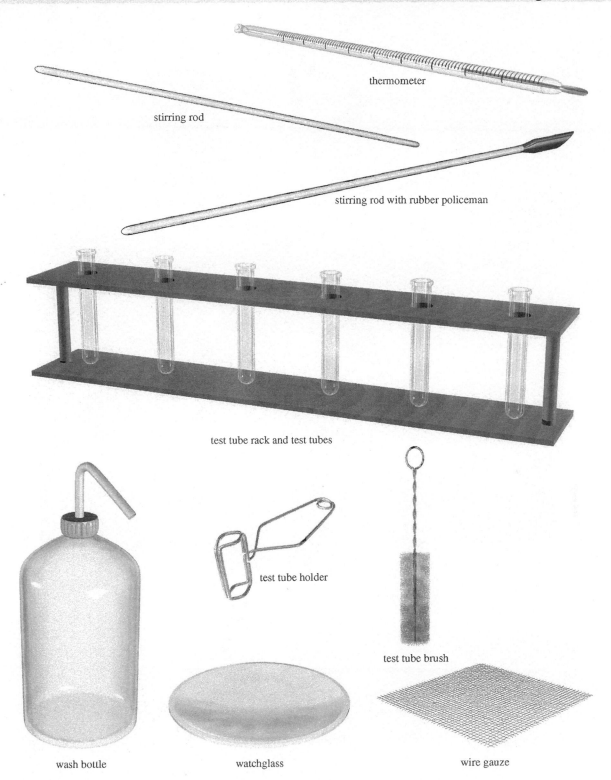

thermometer

stirring rod

stirring rod with rubber policeman

test tube rack and test tubes

test tube holder

test tube brush

wash bottle

watchglass

wire gauze

Introduction to Chemistry

- To gain experience in recording data and explaining observations.
- To develop skill in handling glassware and transferring chemicals.
- To become familiar with basic safety precautions in the laboratory.

DISCUSSION

Chemistry is a science that studies the composition and properties of matter. We can define the term **science** as the methodical exploration of nature and the logical interpretation of the observations. In an **experiment**, scientists gather data and carefully record observations under controlled conditions. After an experiment, scientists formulate a tentative proposal, or **hypothesis**, to explain the data. If additional experiments support the original proposal, a hypothesis may be elevated to a scientific principle, or **theory**. This stepwise procedure is known as the **scientific method** and can be summarized as follows:

Step 1: Perform a planned experiment, make observations, and record data.
Step 2: Analyze the data, and propose a tentative hypothesis to explain the observations.
Step 3: Conduct additional experiments to test the hypothesis. If the evidence supports the initial proposal, the hypothesis may become a theory.

We should note that scientists exercise caution before accepting a theory. Experience has shown that nature reveals her secrets slowly and only after considerable probing. The following example exercise illustrates the scientific method.

Turquoise, a blue mineral, is heated strongly in a test tube for two minutes. A colorless, odorless liquid collects inside the test tube and the blue mineral turns into a white powder.

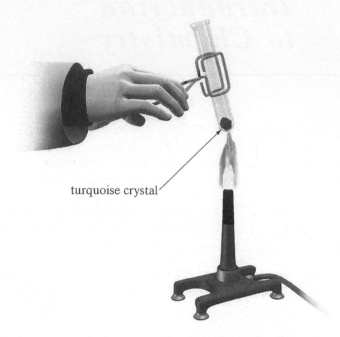

turquoise crystal

Figure 1.1 Heating Blue Turquoise The blue mineral turquoise changes to a white powder and releases a colorless, odorless liquid upon heating.

Observation

- A colorless, odorless liquid collects after strongly heating the test tube.
- The blue mineral changes to a white powder after heating.

Hypothesis

- Water vapor is produced by heating mineral turquoise.
- Turquoise changes to a white powder after losing water.

EQUIPMENT and CHEMICALS

A. Instructor Demonstrations

- tall glass cylinder
- large Erlenmeyer flask + stopper
- glass stirring rod
- 150-mL beaker
- matches
- fire extinguisher
- mortar and pestle
- wash bottle
- evaporating dish

- cupric sulfate solution, 0.1 M $CuSO_4$
- ammonium hydroxide, 6 M NH_4OH
- copper penny (pre-1982 mint date)
- concentrated nitric acid, 16 M HNO_3
- sugar, powdered $C_{12}H_{22}O_{11}$
- concentrated sulfuric acid, 18 M H_2SO_4
- ethyl alcohol, CH_3CH_2OH
- ammonium nitrate, solid NH_4NO_3
- zinc, Zn powder

B. Student Experiments

- 1000-mL Florence flask + stopper and disappearing blue solution
- 13 × 100 mm test tubes (2)
- spatulas (2)
- 250-mL beaker
- 100-mL beaker
- ball-and-stick models

- (10 g glucose in 300 mL 0.5 M KOH + 10 mL of 0.1 g/L methylene blue solution)
- ammonium chloride, solid NH_4Cl
- calcium chloride, solid $CaCl_2$
- iron, Fe nail
- calcium, Ca metal
- copper(II) sulfate solution, 0.1 M $CuSO_4$

PROCEDURE

A. Instructor Demonstrations

The following chemical demonstrations are performed by the Instructor. Students record observations and propose a hypothesis to explain their observation.

1. *Cold Heat.* Add 40 mL of ethyl alcohol to 60 mL of water in a 150-mL beaker. Soak a cotton handkerchief in the alcohol solution, and squeeze out the excess. Hold the handkerchief with crucible tongs, dim the room lights, and ignite.

   Note: The Instructor may wish to point out the location of the fire extinguisher and flammable solvents in the laboratory. Students should try to explain why the cotton handkerchief, soaked in alcohol, does not burn.

2. *Black Foam.* Half fill a 150-mL beaker with household powdered sugar. Add 15 mL of concentrated sulfuric acid, and stir slowly with a glass rod.

   Note: Students should try to identify the black foam. The formula for ordinary powdered sugar is $C_{12}H_{22}O_{11}$.

3. *Copper Smog.* Drop a copper penny (pre-1982 mint date) into a large Erlenmeyer flask. Pour a few milliliters of concentrated nitric acid into the flask to cover the penny, and insert a rubber stopper into the flask. After the penny has stopped reacting, pour the solution into a large beaker of water and observe the color.

   Note: The Instructor should release the gas, NO_2, under a fume hood. Students should try to explain the brown smog and the blue solution.

4. *Here and Gone.* Measure about 100 mL of 0.1 M copper(II) sulfate into a tall glass cylinder. Add about 5 mL of 6 M ammonium hydroxide solution to the cylinder, and observe the reaction. Add an additional 100 mL of 6 M ammonium hydroxide to the cylinder. Observe the reaction, and propose a hypothesis for the observations.

   Note: Students should try to explain the formation of the blue-white solid and its disappearance to form a deep violet solution.

5. *Water Hazard.* Grind about 3 g of ammonium nitrate in a mortar with a pestle. Empty the powder into an evaporating dish; sprinkle fresh zinc dust over the mixture. Deliver a stream of distilled water from a wash bottle onto the chemicals.

   Note: The reaction is exothermic and should be performed with CAUTION under a fume hood. Students should try to explain the intense reaction.

B. Student Experiments

Students perform each of the following as a chemical demonstration. Students will record observations and propose a hypothesis to explain their observation.

1. *Disappearing Blue.* Observe the clear solution in the 1000-mL Florence flask with stopper. Shake the flask once with your thumb firmly holding the stopper. Wait several seconds; repeat the procedure and record your observations.

   Note: Do not discard the blue solution in the Florence flask, as it can be used repeatedly as a chemical demonstration.

2. *Hot and Cold.* Add a small spatula of ammonium chloride into one test tube and a spatula of calcium chloride into a second. Half fill the test tubes with distilled water. Place your hand around the bottom of each test tube, and record your observations.

   Note: Empty the contents of each test tube into the sink, followed by water.

3. *Active and Unreactive.* Half-fill a 250-mL beaker with distilled water. Place an iron nail and a small piece of calcium metal in the water and record your observations.

   Note: Remove the iron nail, and empty the contents of the beaker into the sink.

4. *Copper Nails.* Half-fill a 100-mL beaker with copper sulfate solution. Place the nail in the solution. Wait several minutes, remove the nail, and record your observations.

   Note: Empty the contents of the beaker into the chemical waste container.

5. *Mirror Images.* Given a ball-and-stick model kit, construct the model shown in Figure 1.2. The letter abbreviations on the balls are as follows: B—black, Y—yellow, O—orange, R—red, and G—green.

   Construct a model identical to the first model. On the second model, switch the positions of the red and yellow balls. Can the two models now be superimposed? Are the two models identical? Diagram each model in the Data Table.

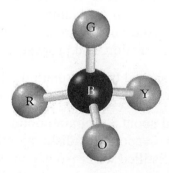

Figure 1.2 Ball-and-Stick Model The illustration shows a molecular model that has a nonidentical mirror image.

NAME _____

SECTION _____

PRELABORATORY ASSIGNMENT*

1. In your own words, define the following terms:

    chemistry

    experiment

    hypothesis

    science

    scientific method

    theory

2. Sketch a simple drawing of the following laboratory equipment:

    Erlenmeyer flask   Florence flask

    beaker   wash bottle

    watchglass   evaporating dish

_____

Answers in Appendix J

3. Which of the following chemicals should be handled carefully in the laboratory: nitric acid, sulfuric acid, ethyl alcohol, ammonium nitrate, distilled water?

4. What should you do if any chemical comes in contact with your skin?

5. What safety precautions must be observed in this experiment?

NAME _____

_____

SECTION _____

DATA TABLE

A. Instructor Demonstrations

    1. *Cold Heat*

Observation	Hypothesis

    2. *Black Foam*

Observation	Hypothesis

    3. *Copper Smog*

Observation	Hypothesis

    4. *Here and Gone*

Observation	Hypothesis

    5. *Water Hazard*

Observation	Hypothesis

B. Student Experiments

1. *Disappearing Blue*

Observation	Hypothesis

2. *Hot and Cold*

Observation	Hypothesis

3. *Active and Unreactive*

Observation	Hypothesis

4. *Copper Nails*

Observation	Hypothesis

5. *Mirror Images*

Observation	Hypothesis

POSTLABORATORY ASSIGNMENT

1. State whether each of the following laboratory safety precautions is *true* or *false*.

   (a) _____   Wear safety goggles while working in the laboratory.

   (b) _____   Wear closed-toe shoes while working in the laboratory.

   (c) _____   Never eat or drink in the laboratory.

   (d) _____   Never taste a chemical in the laboratory.

   (e) _____   Never perform unauthorized experiments.

   (f) _____   Obey the safety precautions for each experiment stated in **Appendix J**.

   (g) _____   Perform experiments under a fume hood that produce gases with an unpleasant odor.

   (h) _____   Never smell a gas or vapor directly; waft the air toward your nose using a cupped hand. (See page 1.)

   (i) _____   Never point the open end of a test tube toward a student when heating a chemical in a test tube. (See page 2.)

   (j) _____   If any chemical contacts your skin or eyes, flush immediately with water, and notify your Instructor.

   (k) _____   Pour acids into water—*not water into acid*—because the heat of solution may cause the acid to spatter.

   (l) _____   Put a drop of glycerin (glycerol) into the hole of a rubber stopper before inserting glass tubing or a thermometer into the stopper.

   (m) _____   Clean up broken glassware immediately.

   (n) _____   Consider all chemicals to be hazardous.

   (o) _____   Do not use a flammable organic liquid near a burner flame.

   (p) _____   Note the location of the fire extinguisher in the laboratory.

   (q) _____   Note the location of the first-aid equipment in the laboratory.

   (r) _____   Notify the Instructor immediately in case of any accident.

2. State whether each of the following laboratory instructions is *true* or *false*.

(a) _____   Do the Prelaboratory Assignment *before* coming to laboratory and check your answers in **Appendix J**.

(b) _____   Record observations directly in the experiment Data Table or notebook; do not record data on loose paper.

(c) _____   Dispose of chemicals in a recycle waste container.

(d) _____   Dispose of organic chemicals in a special waste container.

(e) _____   When an experiment requires water, use distilled water.

(f) _____   Clean glassware with tap water, and rinse with distilled water.

(g) _____   Never place chemicals directly on a balance pan.

(h) _____   Never place hot objects on a balance pan.

(i) _____   Allow porcelain and glassware to cool before moving to the desktop.

(j) _____   Clean your lab station and equipment after completing the experiment.

3. (optional) You have nine pennies; one penny was minted in 1980, and the other eight pennies were minted after 1982. The 1980 penny is 20% heavier because it contains more copper. Assuming the mint dates are illegible, devise a method using the platform balance shown below to determine the heavier 1980 penny in only two weighings.

Instrumental Measurements

- To obtain measurements of length, mass, volume, and temperature.
- To determine the mass and volume of an unknown rectangular solid.
- To gain proficiency in using the following instruments: metric rulers, balances, graduated cylinder, and thermometer.

The **metric system** uses a basic set of units and prefixes. The basic unit of length is the meter, the basic unit of mass is the gram, and the basic unit of volume is the liter. Metric prefixes make these basic units larger or smaller by powers of 10. For example, a kilometer is a thousand times longer than a meter, and a millimeter is a thousand times less than a meter. In the laboratory, the most common unit of length is **centimeter** (symbol **cm**), the most common unit of mass is **gram** (symbol **g**), and the most common unit of volume is **milliliter** (symbol **mL**).

Scientific instruments have evolved to a high state of sensitivity. However, it is not possible to make an exact measurement. The reason is that all instruments possess a degree of **uncertainty**—no matter how sensitive. The uncertainty is indicated by the significant digits in the measurement. For example, a metric ruler may measure length to the nearest tenth of a centimeter (± 0.1 cm). A different metric ruler may measure length to the nearest five hundredths of a centimeter (± 0.05 cm). The measurement with least uncertainty (± 0.05 cm) is more precise.

In this experiment, we will use several instruments. We will make measurements of mass with balances having progressively greater sensitivity. A decigram balance is so named because the uncertainty is one-tenth of a gram (± 0.1 g). The uncertainty of a centigram balance is one-hundredth of a gram (± 0.01 g), and the uncertainty of a milligram balance is one-thousandth of a gram (± 0.001 g).

We will make length measurements using two metric rulers that differ in their uncertainty. METRIC RULER A is calibrated in 1-cm divisions and has an uncertainty of ± 0.1 cm. METRIC RULER B has 0.1-cm subdivisions and an uncertainty of ± 0.05 cm. Thus, METRIC RULER B has less uncertainty than METRIC RULER A. The following examples demonstrate measurement of length utilizing the two different metric rulers.

Example Exercise 2.1 • Measuring Length with Metric Ruler A

A copper rod is measured with the metric ruler shown below. What is the length of the rod?

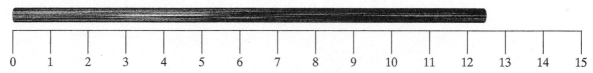

METRIC RULER A   *(Estimate to a tenth of a division ±0.1 cm)*

Solution: Each division represents one centimeter. The end of the rod lies between the 12th and 13th divisions. We can estimate to a tenth of a division (± 0.1 cm). Since the end of the rod lies about five-tenths past 12, we can estimate the length as

$$12 \text{ cm} + 0.5 \text{ cm} = 12.5 \text{ cm}$$

Example Exercise 2.2 • Measuring Length with Metric Ruler B

The same copper rod is measured with the metric ruler shown below. What is the length of the rod?

METRIC RULER B   *(Estimate to a half of a subdivision ±0.05 cm)*

Solution: Note that this ruler is divided into centimeters that are subdivided into tenths of centimeters. The end of the rod lies between the 12th and 13th divisions and between the 5th and 6th subdivisions. Thus, the length is between 12.5 cm and 12.6 cm.

We can estimate the measurement more precisely. A subdivision is too small to divide into ten parts but we can estimate to half of a subdivision (± 0.05 cm). The length is 12 cm + 0.5 cm + 0.05 cm = 12.55 cm.

We can measure the volume of a liquid using a graduated cylinder. If we carefully examine the 100-mL graduated cylinder shown in Figure 2.1, we notice that it is marked in 10-mL intervals, and each interval has ten subdivisions. Therefore, each subdivision equals one milliliter. If we estimate to half of a subdivision, the uncertainty is ± 0.5 mL.

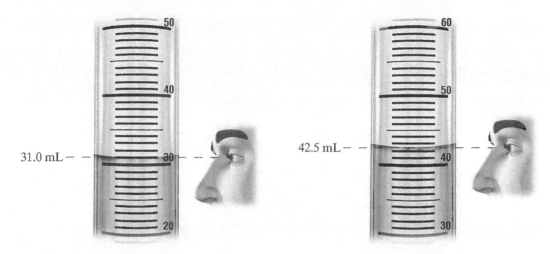

Figure 2.1 Graduated Cylinder Example readings using proper eye position and recording the bottom of the lens to half a subdivision (± 0.5 mL).

We can measure the temperature using a Celsius thermometer. If we examine the thermometer shown in Figure 2.2, we notice that it is marked in 10°C intervals that have ten subdivisions. Thus, each subdivision equals one degree Celsius. If we estimate to half of a subdivision, the temperature measurement has an uncertainty of ± 0.5°C.

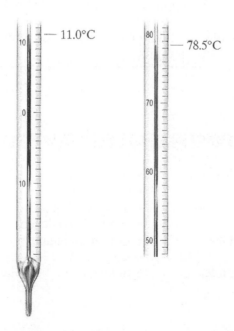

Figure 2.2 Celsius Thermometer Example readings obtained using a Celsius thermometer and recording the top of the liquid to half a subdivision (± 0.5°C).

To test your skill in making metric measurements, you will find the mass and volume of an unknown rectangular solid. The volume of a rectangular solid is calculated from its length, width, and thickness. The following examples will illustrate.

Example Exercise 2.3 • Calculating Volume of a Rectangular Solid

An unknown rectangular solid was measured with METRIC RULER A, which provided the following dimensions: 5.0 cm by 2.5 cm by 1.1 cm. What is the volume of the solid?

Solution: The volume of a rectangular solid equals length times width times thickness.

$$5.0 \text{ cm} \times 2.5 \text{ cm} \times 1.1 \text{ cm} = 13.75 \text{ cm}^3 = 14 \text{ cm}^3$$

When multiplying, the product is limited by the least number of significant digits. In this example, each dimension has two significant digits; thus, the volume is limited to two significant digits.

Example Exercise 2.4 • Calculating Volume of a Rectangular Solid

The unknown rectangular solid was also measured with METRIC RULER B, which gave the following: 5.00 cm by 2.45 cm by 1.15 cm. What is the volume of the solid?

Solution: The volume is once again found by multiplying the three dimensions.

$$5.00 \text{ cm} \times 2.45 \text{ cm} \times 1.15 \text{ cm} = 14.0875 \text{ cm}^3 = 14.1 \text{ cm}^3$$

In this example, each dimension has three significant digits; thus, the volume has three significant digits.

EQUIPMENT and CHEMICALS

- 13 × 100 mm test tubes (3)
- watchglass
- evaporating dish
- 250-mL beaker
- 125-mL Erlenmeyer flask
- crucible and cover
- 100-mL graduated cylinder
- dropper pipet
- 110 °C thermometer
- 150-mL beaker

- ring stand
- wire gauze
- decigram balance
- centigram balance
- milligram balance
- ice
- unknown rectangular solids

A. Length Measurement

1. Measure the length of a 13×100 mm test tube with each of the following: (a) METRIC RULER A, and (b) METRIC RULER B.

   Note: Refer to METRIC RULER A instructions in **Example Exercise 2.1**.
   Refer to METRIC RULER B instructions in **Example Exercise 2.2**.

2. Measure the diameter of a watchglass with each of the following: (a) METRIC RULER A, and (b) METRIC RULER B.

   Note: Refer to METRIC RULER A instructions in **Example Exercise 2.1**.
   Refer to METRIC RULER B instructions in **Example Exercise 2.2**.

3. Measure the diameter of an evaporating dish (not the spout) with each of the following: (a) METRIC RULER A, and (b) METRIC RULER B.

   Note: Refer to METRIC RULER A instructions in **Example Exercise 2.1**.
   Refer to METRIC RULER B instructions in **Example Exercise 2.2**.

B. Mass Measurement

1. Determine the mass of a 250-mL beaker on each of the following balances: (a) decigram balance, (b) centigram balance, and (c) milligram balance.

2. Determine the mass of a 125-mL Erlenmeyer flask on each of the following balances: (a) decigram balance, (b) centigram balance, and (c) milligram balance.

3. Determine the mass of a crucible and cover on each of the following balances: (a) decigram balance, (b) centigram balance, and (c) milligram balance.

   Note: Refer to balance instructions in **Appendices B**, **C**, and **D**.

C. Volume Measurement

1. Fill a 100-mL graduated cylinder with water. Adjust the bottom of the lens to the full mark with a dropper pipet. Record the volume as 100.0 mL.

2. Fill a 13×100 mm test tube with water from the graduated cylinder. Record the new volume in the graduated cylinder (± 0.5 mL).

   Note: Refer to the graduated cylinder instructions in **Figure 2.1**.

3. Fill a second test tube with water. Record the volume in the graduated cylinder.

D. Temperature Measurement

1. Record the temperature in the laboratory using a Celsius thermometer (± 0.5°C).

2. Half fill a 100-mL beaker with ice and water. Hold the thermometer in the ice water, and record the coldest observed temperature (± 0.5°C).

3. Half fill a 150-mL beaker with distilled water. Support the beaker on a ring stand with a wire gauze as shown in Figure 2.4. Heat the water to boiling and shut off the burner. Hold the thermometer in the boiling water and record the hottest temperature (± 0.5°C).

> **Note:** Refer to the laboratory burner instructions in **Appendix A**. You should not let the tip of the thermometer touch the bottom of the glass beaker.

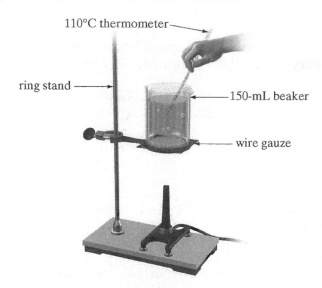

Figure 2.4 Temperature of Boiling Water To avoid breakage, do not allow the thermometer to touch the hot glass beaker.

E. Mass and Volume of an Unknown Solid

1. Obtain a rectangular solid, and record the unknown number in the Data Table. Find the mass of the unknown rectangular solid using each of the following: (a) a decigram balance, (b) a centigram balance, and (c) a milligram balance.

2. Measure the length, width, and thickness of the rectangular solid unknown using METRIC RULER A in Figure 2.3. Calculate the volume.

3. Measure the length, width, and thickness of the rectangular solid unknown using METRIC RULER B in Figure 2.3. Calculate the volume.

NAME _____

SECTION _____

PRELABORATORY ASSIGNMENT*

1. In your own words, define the following terms:

    centimeter (cm)

    gram (g)

    metric system

    milliliter (mL)

    uncertainty

2. State the uncertainty in the following measuring instruments.

    (a) METRIC RULER A _____   (b) METRIC RULER B _____

    (c) decigram balance _____   (d) centigram balance _____

    (e) milligram balance _____   (f) graduated cylinder _____

    (g) thermometer _____

3. Refer to Example Exercises 2.1 and 2.2 and record the measurement indicated by each of the following metric rulers.

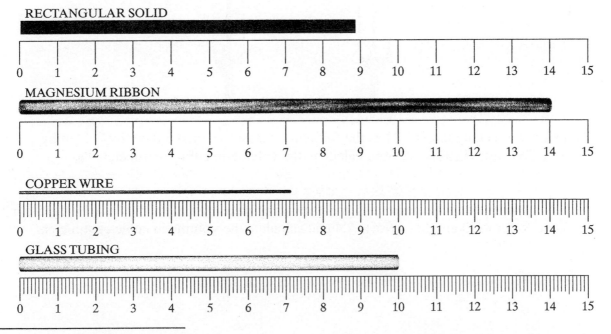

RECTANGULAR SOLID

MAGNESIUM RIBBON

COPPER WIRE

GLASS TUBING

---

* Answers in Appendix J

4. Refer to Figure 2.1 and record the measurement indicated by each graduated cylinder.

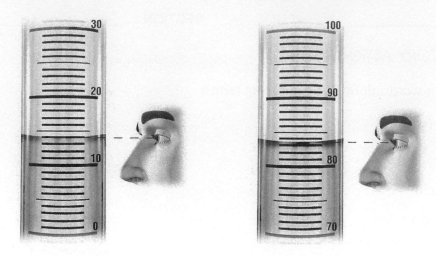

5. Refer to Figure 2.2 and record the measurement indicated by each Celsius thermometer.

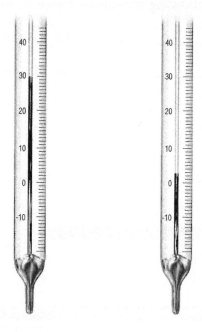

6. An unknown rectangular solid has the following measurements: 5.0 cm by 2.4 cm by 1.3 cm. Refer to Example Exercise 2.3 and calculate the volume in cubic centimeters.

7. An unknown rectangular solid has the following measurements: 5.00 cm by 2.45 cm by 1.25 cm. Refer to Example Exercise 2.4 and calculate the volume in cubic centimeters.

8. What safety precautions must be observed in this experiment?

EXPERIMENT 2   NAME _____

DATE _____   SECTION _____

DATA TABLE

A. Length Measurement

length of a 13 × 100 mm test tube

    METRIC RULER A   _____ cm

    METRIC RULER B   _____ cm

diameter of a watchglass

    METRIC RULER A   _____ cm

    METRIC RULER B   _____ cm

diameter of an evaporating dish

    METRIC RULER A   _____ cm

    METRIC RULER B   _____ cm

B. Mass Measurement

mass of a 250-mL beaker

    decigram balance   _____ g

    centigram balance   _____ g

    milligram balance   _____ g

mass of a 125-mL Erlenmeyer flask

    decigram balance   _____ g

    centigram balance   _____ g

    milligram balance   _____ g

mass of a crucible and cover

    decigram balance   _____ g

    centigram balance   _____ g

    milligram balance   _____ g

C. Volume Measurement

volume of water in a graduated cylinder   _____ mL

volume minus one test tube of water   _____ mL

volume minus two test tubes of water   _____ mL

D. Temperature Measurement

room temperature   _____ °C

melting point temperature of ice   _____ °C

boiling point temperature of water   _____ °C

E. Mass and Volume of an Unknown Solid   **UNKNOWN #** _____

mass of unknown solid

decigram balance   _____ g

centigram balance   _____ g

milligram balance   _____ g

volume of unknown solid (METRIC RULER A)

length of solid   _____ cm

width of solid   _____ cm

thickness of solid   _____ cm

Show the calculation for the volume of the rectangular solid (see Example Exercise 2.3).

_____ cm^3

volume of unknown solid (METRIC RULER B)

length of solid   _____ cm

width of solid   _____ cm

thickness of solid   _____ cm

Show the calculation for the volume of the rectangular solid (see Example Exercise 2.4).

_____ cm^3

NAME _____

SECTION _____

1. State the basic unit and symbol in the metric system for each of the following quantities.

   (a) length   _____   (b) mass   _____

   (c) volume   _____   (d) temperature   _____

2. State a common metric unit and symbol obtained from each of the following instruments.

   (a) metric ruler   _____   (b) balance   _____

   (c) graduated cylinder   _____   (d) thermometer   _____

3. State a common laboratory instrument for measuring each of the following examples.

   (a) diameter of beaker   _____   (b) mass of sugar   _____

   (c) volume of alcohol   _____   (d) temperature of air   _____

4. Select the measurement that is consistent with the uncertainty of each instrument.

   (a) METRIC RULER A: 10 cm, 10.0 cm, 10.00 cm   _____

   (b) METRIC RULER B: 10 cm, 10.0 cm, 10.00 cm   _____

   (c) decigram balance: 10.0 g, 10.00 g, 10.000 g   _____

   (d) centigram balance: 10.0 g, 10.00 g, 10.000 g   _____

   (e) milligram balance: 10.0 g, 10.00 g, 10.000 g   _____

   (f) graduated cylinder: 10 mL, 10.0 mL, 10.00 mL   _____

   (g) Celsius thermometer: 10 °C, 10.0 °C, 10.00 °C   _____

5. State the uncertainty (for example, ± 0.5 cm) in each of the following measurements.

   (a) 10.00 cm   _____   (b) 10.000 g   _____

   (c) 10.0 mL   _____   (d) 10.0 °C   _____

6. State the number of significant digits in each of the following measurements.

   (a) 1.00 cm   _____   (b) 0.05 cm   _____

   (c) 1.000 g   _____   (d) 0.050 g   _____

   (e) 10.0 mL   _____   (f) 50.0 mL   _____

   (g) 110.0 °C   _____   (h) −20.0°C   _____

7. Perform the indicated math operation and round off the answer to the proper significant digits.

   (a)   50.55 g   (b)   50.55 mL
          +   10.050 g   −   10.5   mL

8. Perform the indicated math operation and round off the answer to the proper significant digits.

   (a)   (10.50 cm) (1.50 cm)   (b) $\dfrac{50.50 \text{ cm}^3}{15.0 \text{ cm}^2}$

9. After recording a measurement using an instrument, do you round off experimental data?

   YES / NO   Explain:

10. After multiplying or dividing using a calculator, do you round off numbers in the display?

   YES / NO   Explain:

11. (optional) The original reference kilogram is a solid cylinder made of a platinum–iridium alloy. If the diameter and height are each 3.90 cm, what is the volume in cubic centimeters?

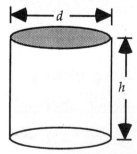

   Note: The volume of a solid cylinder equals $\pi d^2 h/4$, where π is 3.14, d is the diameter, and h is the height.

Density of Liquids and Solids

OBJECTIVES

- To observe the relative densities of some common liquids and solids.
- To determine the densities of water, an unknown liquid, a rubber stopper, and an unknown rectangular solid.
- To determine the thickness of a piece of aluminum foil using the density concept.
- To gain proficiency in performing the following experimental procedures: pipetting a liquid, weighing by difference, and determining a volume by displacement.

DISCUSSION

Density is a physical property of liquids and solids. We can define **density** (symbol d) as the amount of mass in a given volume. To determine the density of a solid experimentally, we must measure the mass of the solid using a balance. To determine the mass of a liquid, we use an indirect technique called **weighing by difference** (Figure 3.1). First, we weigh a flask empty. Second, we add a given volume of liquid into the flask and reweigh. The mass of the liquid is found by subtracting the first and second weighings.

After collecting the experimental data, we can calculate density by dividing the mass by the volume. It is important, however, that we attach the proper units to the calculated value. The density of liquids and solids is usually expressed in grams per milliliter (g/mL) or grams per cubic centimeter (g/cm^3). Since 1 mL = 1 cm^3, the numerical value for density in g/mL and g/cm^3 is identical. For example, the density of water may be expressed as 1.00 g/mL or 1.00 g/cm^3.

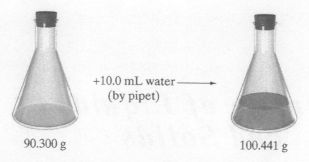

+10.0 mL water ⟶
(by pipet)

90.300 g   100.441 g

Figure 3.1 Weighing by Difference The mass of the liquid is found by the difference in masses: 100.441 g – 90.300 g = 10.141 g.

Example Exercise 3.1 • Density of a Liquid

A 10.0-mL sample of water is pipetted into a flask. The mass of water, 10.141 g, is found after weighing by difference (see Figure 3.1). Calculate the density of water.

Solution: Dividing the mass of water by volume, we have

$$\frac{10.141 \text{ g}}{10.0 \text{ mL}} = 1.01 \text{ g/mL}$$

We round the answer to three significant digits because there are only three digits in the denominator. In this example, the calculated value, 1.01 g/mL, agrees closely with the theoretical value, 1.00 g/mL. The slight discrepancy is due to experimental error.

The volume of an irregular object can be found indirectly from the amount of water it displaces. This technique is called **volume by displacement**. For example, the volume of a rubber stopper can be determined as shown in Figure 3.2. The initial reading of water in the graduated cylinder is observed. The stopper is introduced into the graduated cylinder and the final reading is recorded. The difference between the initial and final readings corresponds to the volume of water displaced. The volume of water displaced is equal to the volume of the rubber stopper.

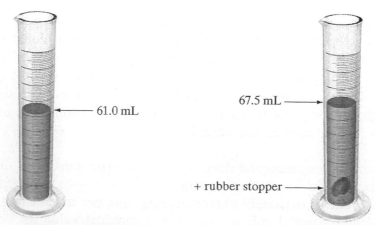

67.5 mL ⟶

⟵ 61.0 mL

+ rubber stopper ⟶

Figure 3.2 Volume by Displacement The volume of the rubber stopper is found by the increase in volume: 67.5 mL – 61.0 mL = 6.5 mL.

Example Exercise 3.2 • Density of a Rubber Stopper

A rubber stopper weighing 8.453 g displaces 6.5 mL of water in a graduated cylinder (Figure 3.2). What is the density of the rubber stopper?

Solution: Dividing the mass of the rubber stopper by its volume, we have

$$\frac{8.453 \text{ g}}{6.5 \text{ mL}} = 1.3 \text{ g/mL}$$

In this example, the volume has only two digits. Thus, the density is limited to two significant digits.

We will also determine the density of a solid. The volume of any solid object with regular dimensions can be found by calculation. For example, the volume of a rectangular solid object is calculated by multiplying its length times its width times its thickness.

Example Exercise 3.3 • Density of a Rectangular Solid

The mass of an unknown rectangular block is 139.443 g. If the block measures 5.00 cm by 2.55 cm by 1.25 cm, what is its density?

Solution: First, we calculate the volume of the rectangular block.

$$5.00 \text{ cm} \times 2.55 \text{ cm} \times 1.25 \text{ cm} = 15.9 \text{ cm}^3$$

Second, we find the density of the unknown rectangular solid.

$$\frac{139.443 \text{ g}}{15.9 \text{ cm}^3} = 8.77 \text{ g/cm}^3$$

The thickness of a sheet of metal foil is too thin to measure with a ruler. However, we can find the thickness indirectly by calculation. Given the mass, length, and width of a metal foil, we can use the density of the metal to calculate the thickness of the foil.

Example Exercise 3.4 • Thickness of an Aluminum Foil

A piece of aluminum foil has a mass of 0.212 g and measures 5.10 cm by 10.25 cm. Given the density of aluminum, 2.70 g/cm^3, calculate the thickness of the foil.

Solution: To calculate the thickness of the foil, we must first find the volume. The volume can be calculated using a density factor as follows.

$$0.212 \text{ g} \times \frac{1 \text{ cm}^3}{2.70 \text{ g}} = 0.0785 \text{ cm}^3$$

The thickness is found after dividing the volume by its length and width.

$$\frac{0.0785 \text{ cm}^3}{(5.10 \text{ cm})(10.25 \text{ cm})} = 0.00150 \text{ cm} \quad (1.50 \times 10^{-3} \text{ cm})$$

A. Instructor Demonstrations

- tall glass cylinder
- methylene chloride (optional)
- hexane

- glass marble
- rubber stopper
- ice
- cork

B–F. Student Experiments

- 125-mL Erlenmeyer flask with stopper
- 150-mL beaker
- 10-mL pipet
- pipet bulb
- 100-mL beaker

- 100-mL graduated cylinder
- #2 rubber stopper
- unknown liquids
- unknown rectangular solids
- aluminum foil, ~10 × 10 cm rectangle

PROCEDURE

A. Instructor Demonstration

1. Half fill a tall glass cylinder with water. Add methylene chloride until two layers are observed. Add hexane until three layers are observed. Record the positions of each layer in the Data Table.

   Note: The Instructor may wish to point out a waste container for chemicals, such as methylene chloride, that should not be poured into the sink.

2. Drop a glass marble into the tall glass cylinder. Record the observation.

3. Drop a rubber stopper into the tall glass cylinder. Record the observation.

4. Drop a piece of ice into the tall glass cylinder. Record the observation.

5. Drop a cork into the tall glass cylinder. Record the observation.

B. Density of Water

1. Weigh a 125-mL Erlenmeyer flask fitted with a rubber stopper.

2. Half fill a 150-mL beaker with distilled water, and then pipet a 10.0 mL sample into the 125-mL flask (see **Appendix E**).

3. Reweigh the flask and stopper, and determine the mass of water by difference.

4. Repeat a second trial for the density of the water.

   Note: It is not necessary to dry the flask between trials because the 10.0-mL sample of water is weighed by difference.

5. Calculate the density of the water for each trial, and report the average value for both trials.

C. Density of an Unknown Liquid

1. Obtain about 25 mL of an unknown liquid in a 100-mL beaker. Record the unknown number in the Data Table.

2. Weigh a 125-mL Erlenmeyer flask fitted with a rubber stopper.

3. Condition a pipet with unknown liquid, and transfer a 10.0-mL sample into the flask.

4. Reweigh the flask and stopper, and determine the mass of liquid by difference.

5. Repeat a second trial for the density of the unknown liquid.

6. Calculate the density of the liquid for each trial, and report the average value for both trials.

D. Density of a Rubber Stopper

1. Weigh a dry #2 rubber stopper.

2. Half fill a 100-mL graduated cylinder with water. Record the water level by observing the bottom of the lens and estimating to ± 0.5 mL.

3. Tilt the graduated cylinder, and let the stopper slowly slide into the water. Record the new level, and calculate the volume by displacement for the stopper.

4. Repeat a second trial for the density of the rubber stopper.

5. Calculate the density of the rubber stopper for each trial, and report the average value for both trials.

E. Density of an Unknown Solid

1. Obtain a rectangular solid, and record the unknown number in the Data Table.

2. Weigh the unknown solid, and record the mass.

3. Measure and record the length, width, and thickness of the unknown rectangular solid, using the metric ruler in Figure 3.3.

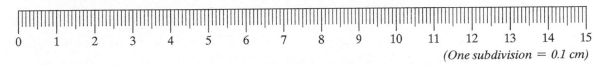

(One subdivision = 0.1 cm)

Figure 3.3 Metric Ruler The uncertainty of the measurement is ± 0.05 cm.

4. Calculate the volume of the unknown rectangular solid.

5. Repeat a second trial for the volume of the unknown solid using a different balance and the metric ruler in Figure 3.3.

F. Thickness of Aluminum Foil

1. Obtain a rectangular piece of aluminum foil.

2. Record the length and width of the foil (see Figure 3.3) in the Data Table.

3. Fold the foil twice. Weigh and record its mass.

4. Calculate the volume and thickness of the aluminum foil ($d = 2.70$ g/cm^3).

PRELABORATORY ASSIGNMENT*

1. In your own words, define the following terms:

   conditioning

   density

   volume by displacement

   weighing by difference

2. Refer to Figure 2.1 and record the volume shown in each of the following graduated cylinders.

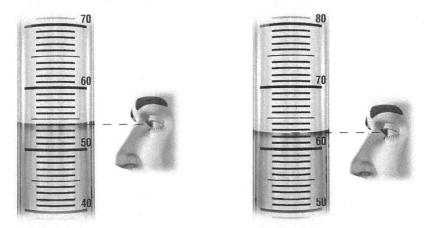

3. Refer to Example Exercise 2.2 and record the length of each rectangular solid.

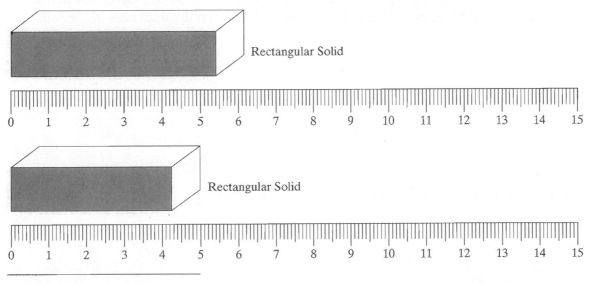

* Answers in Appendix J

4. A 10.0-mL sample of alcohol is pipetted into a flask with stopper. The mass is found by difference to be 7.899 g. Refer to Example Exercise 3.1 and calculate the density of the liquid.

5. A rubber stopper weighing 8.554 g displaces 7.0 mL of water in a graduated cylinder. Refer to Example Exercise 3.2 and calculate the density of the solid rubber stopper.

6 A rectangular titanium solid weighs 152.213 g and measures 7.50 cm by 3.00 cm by 1.50 cm. Refer to Example Exercise 3.3 and calculate the density of the titanium metal.

7. An aluminum foil weighs 0.500 g and measures 10.00 cm by 10.00 cm. Given the density of aluminum is 2.70 g/cm^3, refer to Example Exercise 3.4 and find the thickness of the foil.

8. What safety precautions must be observed in this experiment?

DATA TABLE

A. Instructor Demonstration   *Observations*

      water added to cylinder

      methylene chloride added to water

      hexane added to water

      glass marble added to cylinder

      rubber stopper added to cylinder

      ice added to cylinder

      cork added to cylinder

Diagram of the Tall Glass Cylinder

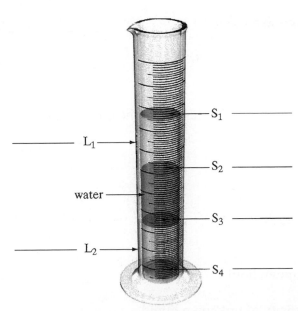

B. Density of Water

mass of flask and stopper + water   _____ g   _____ g

mass of flask and stopper   _____ g   _____ g

mass of water   _____ g   _____ g

volume of water   _____ mL   _____ mL

Show the calculation for the density of water for trial 1 (see Example Exercise 3.1).

Density of water   _____ g/mL   _____ g/mL

Average density of water   _____ g/mL

C. Density of an Unknown Liquid   **UNKNOWN #** _____

mass of flask and stopper + liquid   _____ g   _____ g

mass of flask and stopper   _____ g   _____ g

mass of unknown liquid   _____ g   _____ g

volume of unknown liquid   _____ mL   _____ mL

Show the calculation for the density of the unknown liquid for trial 1.

Density of unknown liquid   _____ g/mL   _____ g/mL

Average density of unknown liquid   _____ g/mL

D. Density of a Rubber Stopper

mass of a rubber stopper _____ g _____ g

final cylinder reading _____ mL _____ mL

initial cylinder reading _____ mL _____ mL

volume of rubber stopper _____ mL _____ mL

Show the calculation of density for the stopper for trial 1 (see Example Exercise 3.2).

Density of rubber stopper _____ g/mL _____ g/mL

Average density of rubber stopper _____ g/mL

E. Density of an Unknown Solid **UNKNOWN #** _____

mass of solid _____ g _____ g

length of solid _____ cm _____ cm

width of solid _____ cm _____ cm

thickness of solid _____ cm _____ cm

Show the calculation for the volume of the unknown for trial 1 (see Example Exercise 3.3).

volume of solid _____ cm^3 _____ cm^3

Show the calculation for the density of the unknown for trial 1 (see Example Exercise 3.3).

Density of rectangular solid _____ g/cm^3 _____ g/cm^3

Average density of the solid _____ g/cm^3

F. Thickness of Aluminum Foil

      length of foil   _____ cm

      width of foil   _____ cm

      mass of foil   _____ g

Show the calculation for the volume of the aluminum foil, $d = 2.70$ g/cm^3 (see Example Exercise 3.4).

      Volume of foil   _____ cm^3

Show the calculation for the thickness of the foil in centimeters.

      Thickness of foil   _____ cm

POSTLABORATORY ASSIGNMENT

1. Ether floats on water, and water floats on mercury, as shown in the following diagram.

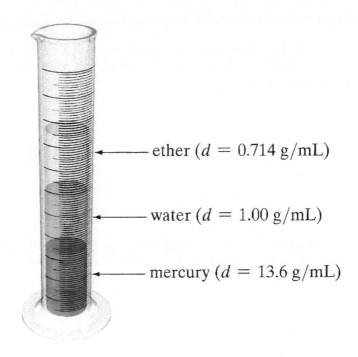

ether ($d = 0.714$ g/mL)

water ($d = 1.00$ g/mL)

mercury ($d = 13.6$ g/mL)

Indicate in the above diagram where each of the following would come to rest after being dropped into the tall glass cylinder.

(a) an aspirin ($d = 1.40$ g/cm^3)

(b) a gold ring ($d = 19.3$ g/cm^3)

(c) a wood chip ($d = 0.40$ g/cm^3)

(d) plastic film ($d = 0.92$ g/cm^3)

2. A 250-mL flask and stopper have a mass of 110.525 g. A 50.0-mL sample of gasoline is pipetted into the flask, and the flask, stopper, and liquid have a mass of 145.028 g. Find the density of the gasoline.

3. A 5-carat diamond has a mass of 1.000 g. If the gemstone displaces a liquid from 1.00 mL to 1.30 mL in a glass buret, what is the density of the diamond?

_____

4. Aluminum foil is often incorrectly referred to as tin foil; however, tin metal is much more dense, 7.28 g/cm^3. Find the thickness of a sheet of tin foil that measures 10.0 cm by 10.0 cm and has a mass of 1.091 g.

_____

5. A thin lead apron is used in dental offices to protect patients from harmful X rays. If the sheet measures 75.0 cm by 55.0 cm by 0.10 cm, and the density of lead is 11.3 g/cm^3, what is the mass of the apron in grams?

_____

6. (optional) If the specific gravity of E85 gasohol is 0.801, what is the mass of 1.00 gallon of the E85 fuel. (Given: 1 gallon = 3.785 liters.)

_____

Freezing Points and Melting Points

- To gain proficiency in constructing a graph and plotting data points.
- To determine the freezing point of a compound from the graph of decreasing temperature versus time.
- To determine the melting points of a known and unknown compound.

A sample of matter can exist in the solid, liquid, or gaseous state. The **physical state** of a substance depends on the temperature and atmospheric pressure. For example, water can exist as solid ice at temperatures below 0°C and as gaseous steam above 100°C.

A **change of state** occurs when there is sufficient heat energy for individual molecules to overcome their attraction for each other. For example, when ice is converted to water, the water molecules in the ice crystal acquire enough energy to become free of each other and move around. Conversely, when water cools to ice, the water molecules lose energy and can no longer move about. Thus, a solid is composed of fixed particles and a liquid has mobile particles.

At the temperature where a liquid changes to a solid, two physical states are present simultaneously. This temperature is referred to as the **freezing point**. Conversely, if a solid changes to a liquid, it is called the **melting point**. Theoretically, the freezing and melting points of a substance occur at the same temperature.

In this experiment, we will melt paradichlorobenzene and then allow the liquid to cool to a solid. We will record the temperature/time relationship, plot the data, and graph a cooling curve. The temperature remains constant as the liquid solidifies. Figure 4.1 shows a typical cooling curve.

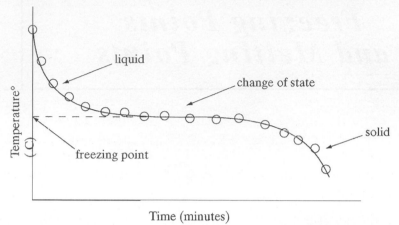

Time (minutes)

Figure 4.1 Cooling Curve As a liquid cools, it changes state from a liquid to a solid. The freezing point corresponds to the flat plateau portion of the curve.

As the compound cools, crystals begin to form. After a few minutes the crystals become a solid mass as the liquid changes to a solid. We will plot temperature on the vertical axis, which is called the **ordinate**. We will plot time on the horizontal axis, which is referred to as the **abscissa**. The freezing point of the compound is the temperature corresponding to the flat plateau. The apparatus for determining the cooling curve is shown in Figure 4.2.

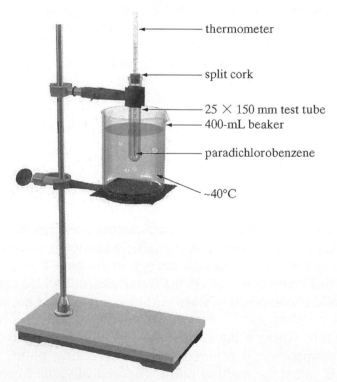

Figure 4.2 Change of State Apparatus The melted paradichlorobenzene is inside a test tube, which in turn, is placed in a beaker of water at ~40°C.

In the second procedure of this experiment a melting point is determined. A small sample of compound is rapidly heated until it is observed to liquefy. The temperature range over which the compound melts is recorded; for example, 65–75°C. A second trial is repeated for greater accuracy. The waterbath is heated rapidly to 60°C and then slowly until the compound melts. This second trial should produce an accurate melting point with a 1–2°C range; for example 69.5–71.0°C.

EQUIPMENT and CHEMICALS

- wire gauze
- mortar and pestle
- 110°C thermometer with split cork
- 400-mL beaker
- 25 × 150 mm test tube containing 20 g of paradichlorobenzene

- 50 cm of 6-mm glass tubing
- capillary tubes
- rubber bands
- biphenyl (diphenyl)
- melting point unknowns

PROCEDURE

A. Cooling Curve and Freezing Point

The Instructor may wish to have students work in pairs. One student should set up the apparatus and record data while the other student heats the paradichlorobenzene and later observes temperature readings.

1. Set up the apparatus as shown in Figure 4.2. Add 300 mL of distilled water to the 400-mL beaker. Heat the water to 40°C, and shut off the burner.

2. Obtain a test tube containing paradichlorobenzene and immerse the test tube in a waterbath of boiling water. After the compound has melted, insert a thermometer into the test tube and continue heating until the liquid is well above 65°C.

    Note: It may be convenient for the Instructor to provide a large, hot waterbath to heat the test tubes containing paradichlorobenzene.

3. Transfer the test tube and thermometer into the 400-mL beaker of water at 40°C. Support the test tube with a utility clamp, and hold the thermometer using a split cork as shown in Figure 4.2.

4. Begin recording thermometer readings when the temperature drops to 65.0°C. Continue recording the temperature (± 0.5°C) every 30 seconds for ten minutes.

5. Plot the temperature/time data on the graph paper provided. Circle each point, and draw a smooth cooling curve. Extend a dashed line from the flat portion of the curve to the vertical axis in order to determine the freezing point of the compound.

    Note: If the thermometer is frozen in solid paradichlorobenzene, **do not attempt to pull out the thermometer**. Return the test tube to the hot waterbath and allow the solid to melt; then remove the thermometer. Do not pour out the liquid paradichlorobenzene, as the compound is used for repeated trials.

B. Melting Point of an Unknown

1. Seal one end of a capillary tube with a burner flame. Let the tube cool, and then dab the open end into a small sample of biphenyl. Invert the capillary and lightly tap the sealed end to pack the sample. Repeat this process until a 5-mm sample is packed at the sealed end of the capillary.

   Note: If the biphenyl crystals are large, grind the crystals using a mortar and pestle. To pack the crystals, drop the sealed end of the capillary through a long piece of 6-mm glass tubing onto the lab bench.

2. Set up an apparatus as shown in Figure 4.3. Add 300 mL of distilled water into the 400-mL beaker. Attach the capillary at the end of the thermometer with a rubber band, and place in the beaker.

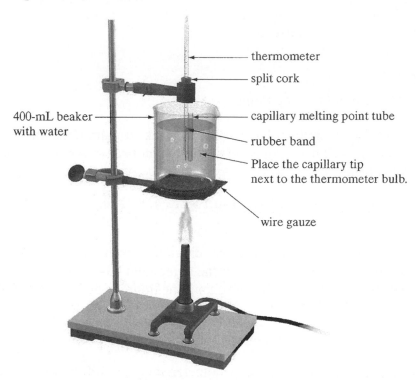

thermometer
split cork
400-mL beaker with water
capillary melting point tube
rubber band
Place the capillary tip next to the thermometer bulb.
wire gauze

Figure 4.3 Melting Point Apparatus The melting point is recorded when the solid melts to a liquid and appears clear in the capillary tube.

3. Rapidly heat the water in the beaker until the biphenyl melts. Observe the approximate melting point ($\pm 1°C$), and record the range of temperature in the Data Table.

4. Prepare another capillary tube and heat rapidly until the temperature is within 10°C of the melting point. Then slowly continue to heat in order to determine the melting point accurately. Record the melting point range ($\pm 0.5°C$) from the first sign of melting until the compound has completely melted. The reference value is given in the Data Table for comparison.

5. Obtain an unknown compound, and record the number. Determine the melting point for the unknown as above.

NAME _____

DATE _____

SECTION _____

1. In your own words, define the following terms:

    abscissa

    change of state

    freezing point

    melting point

    ordinate

    origin

    physical state

2. Why is distilled water used in a hot waterbath?

3. When the test tube with hot liquid paradichlorobenzene is placed in the beaker of water to cool, what is the initial temperature of the water in the beaker?

4. When the test tube with hot liquid paradichlorobenzene is placed in the beaker of water, what is the initial temperature of the paradichlorobenzene in the test tube?

5. When the test tube with hot liquid paradichlorobenzene is placed in the beaker of water, what is the initial *recorded* temperature reading?

6. The freezing point of paradichlorobenzene corresponds to which point on the cooling curve?

Answers in Appendix J

7. After the liquid paradichlorobenzene freezes, how is the thermometer removed from the frozen solid?

8. In determining a melting point of a compound, why are two trials performed?

9. While performing the first melting point trial, a compound begins to melt at 65°C and liquefies completely at 75°C. Report the approximate melting point.

10. While performing the second melting point trial, a compound begins to melt at 68.0°C and liquefies completely at 69.5°C. Report the precise melting point.

11. A solid compound in a capillary tube is placed in the waterbath and appears to liquefy before heating. Give two possible explanations for the problem.

   (1)

   (2)

12. What safety precautions must be observed in this experiment?

DATE _____

DATA TABLE

A. Cooling Curve and Freezing Point

Temperature	Time	Observation
65.0°C	0:00	liquid
	0:30	
	1:00	
	1:30	
	2:00	
	2:30	
	3:00	
	3:30	
	4:00	
	4:30	
	5:00	
	5:30	
	6:00	
	6:30	
	7:00	
	7:30	
	8:00	
	8:30	
	9:00	
	9:30	
	10:00	

B. Melting Point of an Unknown

	Rapid Trial	Trial 2
Mp of biphenyl (69–71°C)	_____ °C	_____ °C
Mp of **UNKNOWN #** _____	_____ °C	_____ °C

Cooling Curve — Trial 2

Temperature	Time	Observation
65.0°C	0:00	liquid
	0:30	
	1:00	
	1:30	
	2:00	
	2:30	
	3:00	
	3:30	
	4:00	
	4:30	
	5:00	
	5:30	
	6:00	
	6:30	
	7:00	
	7:30	
	8:00	
	8:30	
	9:00	
	9:30	
	10:00	

A. Cooling Curve — Trial 1   Freezing Point:_____° C

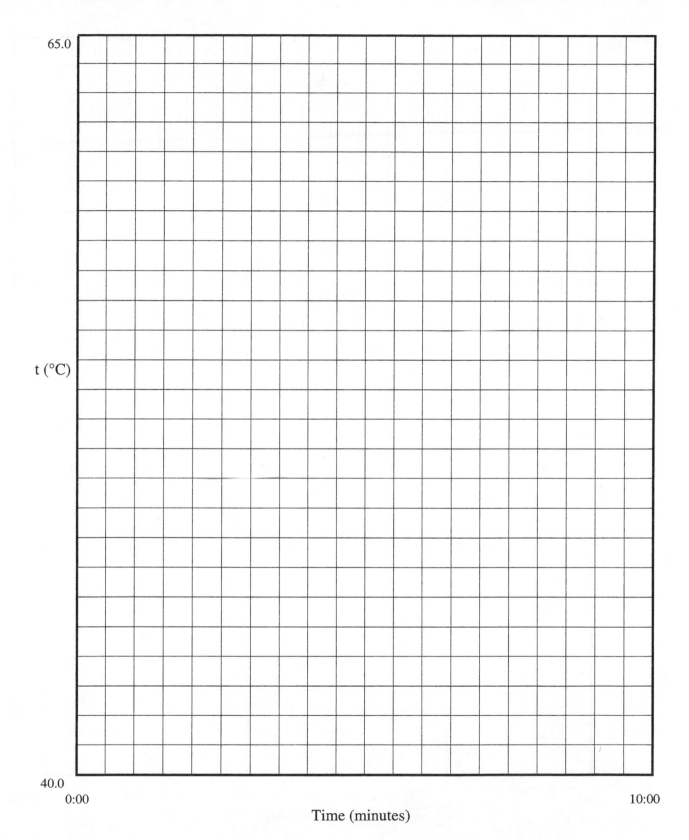

t (°C)

65.0

40.0

0:00   10:00

Time (minutes)

Freezing Point:_____°C

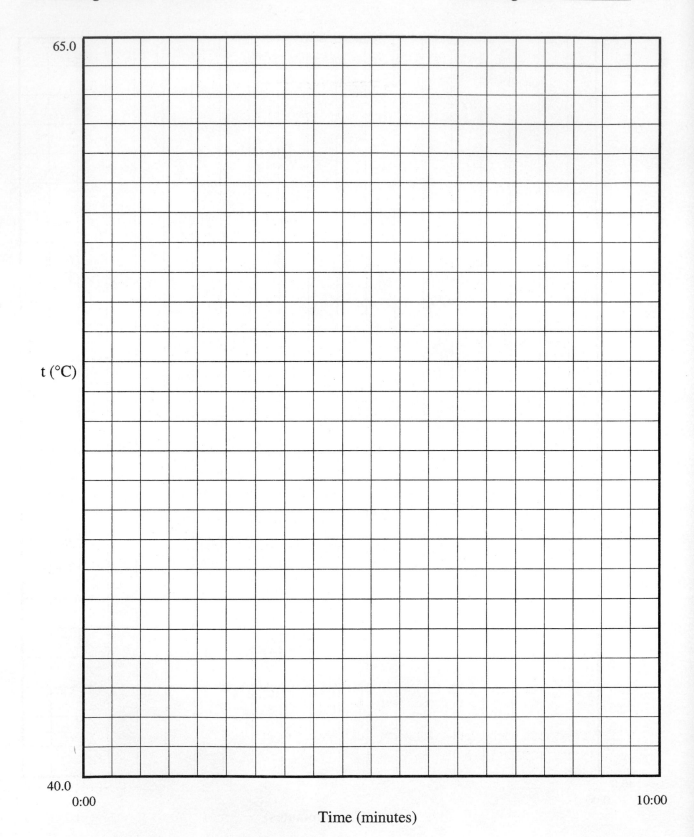

65.0

t (°C)

40.0

0:00

Time (minutes)

10:00

NAME _____

DATE _____

SECTION _____

1. Naphthalene is a compound used in closets to destroy moth larva and protect clothes. Use the following data for naphthalene to graph the cooling curve.

Temperature (°C)	Time (minutes)
83.0	0:00
81.5	0:30
81.0	1:00
80.5	1:30
80.5	2:00
80.5	2:30
80.5	3:00
80.5	3:30
80.5	4:00
80.0	4:30
79.5	5:00

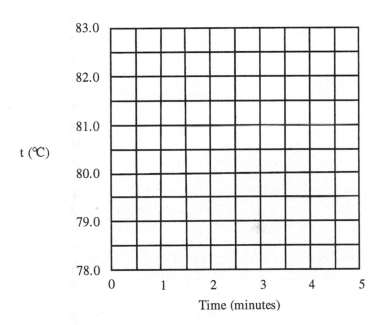

From the graph, estimate the freezing point of naphthalene ± 0.5 °C. _____

2. Ethylene glycol, antifreeze, freezes to a solid at −12 °C. What is the freezing point on the Kelvin and Fahrenheit scales?

_____

_____

3. The following graph shows a heating curve for iodine, I_2, which undergoes sublimation from dark-violet crystals to a purple vapor:

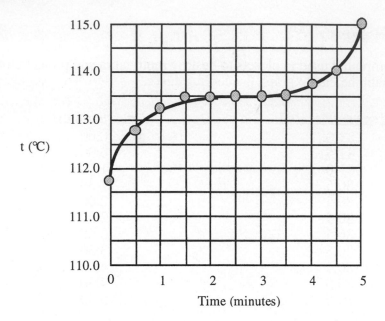

From the graph, estimate the sublimation temperature ± 0.5 °C.   _____

4. Dry ice, solid CO_2, undergoes sublimation from a white solid to a gas at −78 °C. What is the sublimation temperature on the Kelvin and Fahrenheit scales?

_____

_____

5. (optional) Using the *Handbook of Chemistry and Physics, Physical Constants of Inorganic Compounds*, find the melting points (°C) of the following.

   (a) sodium metal, Na   _____

   (b) sodium chloride, NaCl   _____

Physical Properties and Chemical Properties

- To observe the appearance of several metals and nonmetals.
- To determine the boiling points of methyl alcohol and an unknown liquid.
- To determine whether a solid is soluble or insoluble in water.
- To determine whether a liquid is soluble or insoluble in water.
- To determine whether a substance is undergoing a physical or chemical change.
- To gain experience in determining a boiling point and observing test tube reactions.

Chemists classify matter according to its physical and chemical properties. Matter can be classified as a mixture or a pure substance, depending upon its properties. A **heterogeneous mixture** has physical and chemical properties that vary within the sample. For example, combining sugar and salt gives a heterogeneous mixture because the properties of sugar and salt are different.

A **homogeneous mixture** has constant properties although the properties can vary from sample to sample. A homogeneous mixture may be a gaseous mixture, a solution, or an alloy. Examples include air, seawater, and brass, which is an alloy of the metals copper and zinc.

A pure **substance** is either an **element** or a **compound**; all substances have constant and predictable properties. Examples include the compound sodium chloride, as well as the elements sodium metal and chlorine gas.

Sugar is a compound that contains the elements carbon, hydrogen, and oxygen. When heated, sugar decomposes into black carbon and water vapor. When electricity is passed through water, it decomposes into elemental hydrogen and oxygen gases. Although hydrogen and oxygen are both colorless, odorless gases, they differ in their other physical and chemical properties. Figure 5.1 illustrates the overall relationship for the classification of matter.

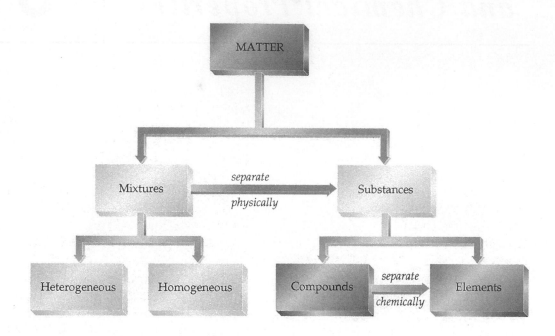

Figure 5.1 Classification of Matter   Matter is either a mixture or a pure substance. The properties of a *heterogeneous mixture* vary within the sample, but the properties of a *homogeneous mixture* are constant. A substance is either a *compound* or an *element*, and the properties of a pure substance are predictable and constant.

A **physical property** refers to a characteristic that can be observed without changing the composition of the substance. A partial list of physical properties include: physical state (solid, liquid, gas), density, malleability, ductility, melting point, boiling point, and solubility in water. A **chemical property** refers to a property that can only be observed during a chemical reaction. The chemical properties of oxygen gas include its ability to react with most metals and nonmetals. On the other hand, helium is an inert gas and its chemical properties reveal that it does not react with other elements.

In this experiment, we will observe a **physical change** as a substance undergoes a change in physical state, a temporary change in color, or a simple change in volume when two solutions are added together. We will observe a **chemical change** when a substance releases a gas, undergoes a permanent change in color, or forms an insoluble substance when two solutions are added together. An antacid tablet fizzing in water, a banana changing color from green to yellow,

In this experiment, we will observe a **physical change** as a substance undergoes a change in physical state, a temporary change in color, or a simple change in volume when two solutions are added together. We will observe a **chemical change** when a substance releases a gas, undergoes a permanent change in color, or forms an insoluble substance when two solutions are added together. An antacid tablet fizzing in water, a banana changing color from green to yellow, and the formation of an insoluble "bathtub ring," are all familiar and practical examples of a chemical change.

EQUIPMENT and CHEMICALS

- ring stand
- wire gauze
- 400-mL beaker
- 16 × 150 mm test tube
- boiling chips
- 110°C thermometer
- split cork
- 13 × 100 mm test tubes (6) and test tube rack
- crucible tongs
- 250-mL beaker
- evaporating dish
- test tube brush
- test tube holder
- wash bottle with distilled water

- small vials with samples of cobalt, hydrogen, magnesium, manganese, neon, oxygen, silicon, sulfur, tin, zinc
- methyl alcohol, CH_3OH
- boiling point unknowns
- copper(II) sulfate crystals, $CuSO_4 \cdot 5H_2O$
- calcium carbonate crystals, $CaCO_3$
- amyl alcohol, $C_5H_{11}OH$
- copper wire, heavy gauge Cu
- iodine, solid crystals I_2
- ammonium bicarbonate, solid NH_4HCO_3
- potassium bicarbonate, solid $KHCO_3$
- sodium carbonate solution, 0.5 M Na_2CO_3
- sodium sulfate solution, 0.1 M Na_2SO_4
- dilute hydrochloric acid, 6 M HCl
- calcium nitrate solution, 0.1 M $Ca(NO_3)_2$
- copper(II) nitrate solution, 0.1 M $Cu(NO_3)_2$
- ammonium hydroxide solution, 6 M NII_4OH

PROCEDURE

A. Physical Properties

1. *Physical State and Color*

   Observe vials of the following elements and record your observations in the Data Table. Classify the element as a metal, a nonmetal, or a semimetal.

   (a) cobalt

   (c) magnesium

   (e) neon

   (g) silicon

   (i) tin

   (b) hydrogen

   (d) manganese

   (f) oxygen

   (h) sulfur

   (j) zinc

2. *Boiling Point*

(a) Place a 400-mL beaker on a wire gauze, and support on a ring stand. Add 300 mL of distilled water to the beaker, bring to a boil, and shut off the burner. Put about 20 drops of methyl alcohol into a 16 × 150 mm test tube. Add a boiling chip and place the test tube in the beaker of hot water. Suspend a thermometer about 1 cm above the liquid. After the alcohol begins to boil in the test tube, record the boiling point temperature (± 0.5°C) when alcohol drips from the tip of the thermometer every few seconds (Figure 5.2).

Caution: Methyl alcohol is flammable, and the vapors must be kept away from the flame of a laboratory burner.

Alternate Apparatus: If hotplates are available, the Instructor may wish to use the alternate apparatus shown in Figure 5.2a.

(b) Record the number of an unknown liquid, and determine the boiling point of the liquid (± 0.5°C) as above.

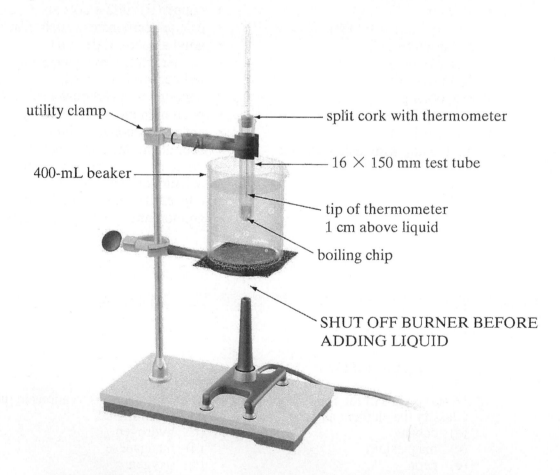

utility clamp

split cork with thermometer

400-mL beaker

16 × 150 mm test tube

tip of thermometer
1 cm above liquid

boiling chip

SHUT OFF BURNER BEFORE
ADDING LIQUID

Figure 5.2 Boiling Point Apparatus The boiling point is recorded when the alcohol vapor condenses to a liquid on the tip of the thermometer, and begins to drip every 2 or 3 seconds.

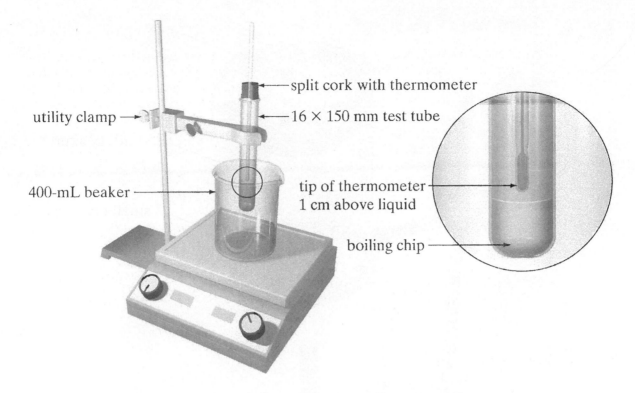

split cork with thermometer

utility clamp →

16 × 150 mm test tube

400-mL beaker —

tip of thermometer
1 cm above liquid

boiling chip —

Figure 5.2a Boiling Point Apparatus Set up the apparatus as shown, turn on the
heat until the water in the beaker begins to boil, then turn off the hotplate.

3. *Solubility of a Solid in Water*

   Add 20 drops of distilled water into two test tubes. Drop a copper sulfate crystal into
   one test tube, and a calcium carbonate crystal into the other. Shake the test tubes briefly
   to observe solubility. State whether each solid is *soluble* or *insoluble* in water.

4. *Solubility of a Liquid in Water*

   Add 20 drops of distilled water in two test tubes. Add a few drops of methyl alcohol to
   one test tube, and amyl alcohol to the other. Shake the test tubes briefly to mix the
   liquids. State whether each liquid is *soluble* or *insoluble* in water.

B. Chemical Properties

   1. *Reactions of Elements*

      (a) Inspect a 5-cm piece of copper wire. Hold the wire with crucible tongs, and heat
          the wire until it glows red. Allow the wire to cool and inspect once again. Classify
          your observation as a *physical change* or *chemical change*.

      (b) Put 3 small crystals of iodine in a dry 250-mL beaker. Cover the beaker with an
          evaporating dish and place ice in the dish (Figure 5.3). Support the beaker on a
          ring stand, and heat the iodine slowly until all the crystals vaporize and the vapor
          deposits on the bottom of the evaporating dish. Classify your observation as a
          physical change or *chemical change*.

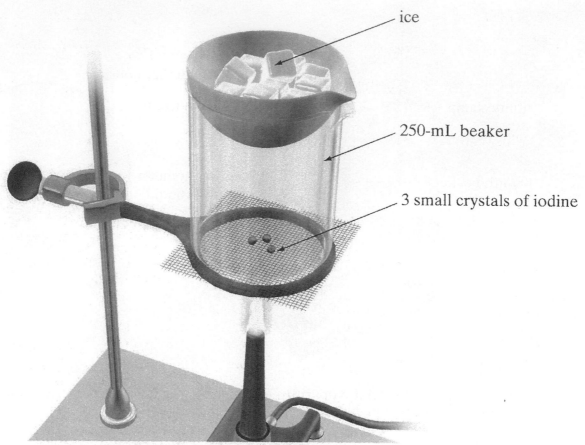

ice

250-mL beaker

3 small crystals of iodine

Figure 5.3 Apparatus for Sublimation/Deposition   Three small crystals of iodine are placed in the beaker and heated gently. The iodine crystals undergo sublimation to a vapor, which deposits on the bottom of the evaporating dish.

2. *Reactions of Compounds*

   (a) Put about 1 g of ammonium bicarbonate into a dry test tube. Use a test tube holder and heat the test tube gently with a cool flame and note any changes including odor. Classify your observation as a *physical change* or *chemical change*.

   Caution: When heating the contents of a test tube, point the open end in a safe direction (see *Safety Precautions*, page 2).

   (b) Put about 1 g of potassium bicarbonate into a dry test tube. Heat the test tube gently with a cool flame and record any changes. Classify your observation as a *physical change* or *chemical change*.

3. *Reactions of Solutions*

   (a) Add 20 drops of sodium carbonate, and 20 drops of sodium sulfate into separate test tubes. Add 20 drops of dilute hydrochloric acid to each test tube, and record any changes. Classify your observation as a *physical change* or *chemical change*.

   (b) Add 20 drops of calcium nitrate, and 20 drops of copper(II) nitrate into separate test tubes. Add 20 drops of dilute ammonium hydroxide to each test tube, and note any changes. Classify your observation as a *physical change* or *chemical change*.

NAME _____

SECTION _____

PRELABORATORY ASSIGNMENT*

1. In your own words, define the following terms:

    chemical change

    chemical property

    compound

    element

    heterogeneous mixture

    homogeneous mixture

    physical change

    physical property

    substance

2. Classify the following characteristics as a physical (*phys*) or chemical (*chem*) property.

    (a) appearance   (b) physical state

    (c) boiling point   (d) hardness

    (e) density   (f) ductility

    (g) malleability   (h) conductivity

    (i) solubility   (j) reactivity

---

* Answers in Appendix J

3. Classify the following observations as a physical (*phys*) or chemical (*chem*) change.

    (a)   wax melting

    (b)   alcohol vaporizing

    (c)   kerosene burning

    (d)   salt dissolving

    (e)   sugar crystallizing

    (f)   antacid bubbling in water

    (g)   fire releasing heat

    (h)   fireworks releasing light

    (i)   fruit changing color

    (j)   sour milk forming insoluble solid

4. What is the purpose of the boiling chip when determining the boiling point of a liquid?

5. What experimental observations suggest a chemical change has taken place?

6. What experimental observations suggest a gas is being released?

7. What safety precautions must be observed in this experiment?

DATE _____

DATA TABLE

A. Physical Properties

   1. *Physical State and Color*

Element	Symbol	Physical State	Color	Metal/Nonmetal
cobalt	_____	_____	_____	_____
hydrogen	_____	_____	_____	_____
magnesium	_____	_____	_____	_____
manganese	_____	_____	_____	_____
neon	_____	_____	_____	_____
oxygen	_____	_____	_____	_____
silicon	_____	_____	_____	_____
sulfur	_____	_____	_____	_____
tin	_____	_____	_____	_____
zinc	_____	_____	_____	_____

   2. *Boiling Point*

      Bp of methyl alcohol (65.0°C)   _____ °C

      Bp of **UNKNOWN #_____**   _____ °C

   3. *Solubility of a Solid in Water*

      copper sulfate crystal and water   _____

      calcium carbonate crystal and water   _____

   4. *Solubility of a Liquid in Water*

      methyl alcohol and water   _____

      amyl alcohol and water   _____

B. Chemical Properties

Procedure	Observation	Physical Change or Chemical Change

1. Reactions of Elements

Procedure	Observation	Physical Change or Chemical Change
(a) copper wire + heat		
(b) iodine crystal + heat		

2. Reactions of Compounds

Procedure	Observation	Physical Change or Chemical Change
(a) ammonium bicarbonate + heat		
(b) potassium bicarbonate + heat		

3. Reactions of Solutions

Procedure	Observation	Physical Change or Chemical Change
(a) sodium carbonate + hydrochloric acid		
sodium sulfate + hydrochloric acid		
(b) calcium nitrate + ammonium hydroxide		
copper(II) nitrate + ammonium hydroxide		

NAME _____

SECTION _____

POSTLABORATORY ASSIGNMENT

1. State whether the following properties are typical of an element on the *left side* or *right side* of the periodic table.

   (a)   shiny, metallic solid   _____   (b)   low melting point   _____

   (c)   malleable, ductile solid   _____   (d)   reacts with metals   _____

2. Classify the following as a physical property (*phys*) or a chemical property (*chem*).

   (a)   Sodium metal in chlorine gas releases energy.   _____

   (b)   Sodium metal has a metallic silver luster.   _____

   (c)   Sodium metal in water produces hydrogen gas.   _____

   (d)   Sodium metal melts to a liquid at 98°C.   _____

   (e)   Sodium metal and oxygen form sodium oxide.   _____

   (f)   Sodium metal conducts heat and electricity.   _____

3. Classify the following as a physical change (*phys*) or a chemical change (*chem*).

   (a)   Sodium metal exposed to air turns dark gray.   _____

   (b)   Steam in a test tube condenses to moisture.   _____

   (c)   Baking soda in vinegar releases bubbles.   _____

   (d)   Water dissolves in vinegar to give a solution.   _____

   (e)   Tap water and soap form an insoluble deposit.   _____

   (f)   Solid "dry ice" slowly disappears completely.   _____

4. Classify each of the following as an example of an *element, compound, homogeneous mixture,* or *heterogeneous mixture.*

   (a)   silver, Ag   _____   (b)   silver chloride, AgCl   _____

   (c)   sterling silver alloy   _____   (d)   silver in quartz rock   _____

5. Using the *Handbook of Chemistry and Physics*, *The Elements*, research the description, specific gravity, melting point, and boiling point, for the following elements.

Name, Symbol	Description	Specific Gravity	Melting Point, °C	Boiling Point, °C
Cobalt, Co	_____	_____	_____	_____
Magnesium, Mg	_____	_____	_____	_____
Manganese, Mn	_____	_____	_____	_____
Silicon, Si	_____	_____	_____	_____
Sulfur, S (rhombic)	_____	_____	_____	_____
Zinc, Zn	_____	_____	_____	_____

6. (optional) Using the online reference Wikipedia at www.wikipedia.org, state the appearance, density, melting point, and boiling point, for the following elements.

Internet Search	Appearance	Density, g/cm^3	Melting Point, °C	Boiling Point, °C
Cobalt	_____	_____	_____	_____
Magnesium	_____	_____	_____	_____
Manganese	_____	_____	_____	_____
Silicon	_____	_____	_____	_____
Sulfur	_____	_____	_____	_____
Zinc	_____	_____	_____	_____

"Atomic Fingerprints"

- To distinguish between a continuous spectrum and a line spectrum.
- To compare observed and calculated lines in the hydrogen spectrum.
- To identify unknown elements in a fluorescent light by "atomic fingerprints."
- To become proficient in using a hand spectroscope.

DISCUSSION

In 1900, Max Planck, a German professor of physics, proposed a revolutionary concept. The concept was quite simple: since matter consists of particles, energy may also consist of particles. This concept was quite revolutionary because light energy had previously been considered a continuous wave and not a collection of discrete particles. We call a particle of matter an atom, and we can refer to a particle of light energy as a **photon**.

Although the terms light and radiant energy are often used interchangeably, **light** usually refers to radiant energy that is visible, rather than invisible radiation such as infrared or ultraviolet. Light travels as a wave, and the **wavelength** is the crest-to-crest distance to complete one cycle. The **frequency** is the number of wave cycles that occur in one second. Low-energy light has a long wavelength and a low frequency. Conversely, high-energy light has a short wavelength and a high frequency.

Radiant Energy Spectrum

When we observe white light, we see the effect of several combined colors of light. When white light passes through a glass prism, however, it separates into six primary colors: red, orange, yellow, green, blue, and violet. This is illustrated in Figure 6.1.

A rainbow is a natural phenomenon that results from sunlight passing through raindrops, which act as miniature prisms to separate sunlight into various bands of color. We can record the wavelength of light in nanometers, where a **nanometer** (**nm**) is one-billionth of a meter. The range of visible light, violet to red, is usually considered 400–700 nm.

We see in Figure 6.2 that the **visible spectrum** is only a small portion of the radiant energy spectrum. The entire spectrum includes X ray, ultraviolet, visible, infrared, and microwave radiation. Our eyes can detect light only in the visible spectrum and not in other regions. That is, the wavelengths of ultraviolet light are too short to be seen by the human eye, and the wavelengths of infrared light are too long to be visible.

Bohr Model of the Atom

In 1913, Niels Bohr proposed that electrons travel in circular orbits around the nucleus, much as planets travel in orbits around the Sun. The Bohr model was a beautiful theory of electrons in atoms. However, no one knew if the model was correct because there was no experimental evidence for the theory. Coincidentally, Bohr received a paper on emitted light from electrically excited hydrogen gas. He noted that hydrogen gas does not emit a **continuous spectrum**, but rather discrete lines. The three most prominent lines are violet, blue-green, and red.

We can describe the experiment as follows. First, hydrogen gas is sealed in a glass tube. Then, an electrical voltage is applied to energize the hydrogen gas. An instant later, the excited hydrogen atoms release energy in the form of light. When the emitted light passes through a prism, it separates into narrow bands of light. This collection of narrow bands of emitted light is called a **line spectrum**. Figure 6.3 illustrates the line spectrum of hydrogen.

As Bohr considered the emission spectrum of hydrogen, he realized that he had powerful experimental evidence for his model of the atom. His concept of energy levels was supported by the line spectrum of hydrogen. He reasoned as follows: When hydrogen is energized using a gas discharge tube, electrons in hydrogen atoms jump to a higher energy level. For example, electrons may jump from the first level to the second, third, fourth, or fifth level. Next, excited electrons lose energy by dropping to a lower level closer to the nucleus.

When an electron drops to a lower energy level, it loses a specific amount of energy that corresponds to the energy of one photon of light. That is, the emitted photon of light has the same amount of energy as that lost by the electron when it drops from a higher to lower energy level. Thus, the line spectrum of hydrogen gas supports Bohr's model of the atom *experimentally*. However, it was Planck's idea that light is composed of particles that supported Bohr's model of the atom *theoretically*. Figure 6.4 shows the line spectrum of hydrogen and the corresponding electron energy levels.

Further study of emission spectra revealed that each element produces a unique set of spectral lines. For this reason, the line spectrum of a given element is sometimes referred to as an **"atomic fingerprint."**

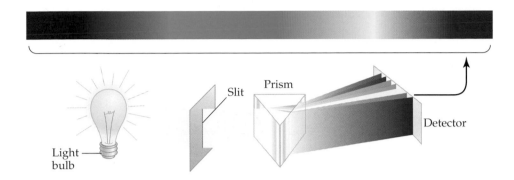

Figure 6.1 White Light Passing Through a Prism White light produces a rainbow of colors when it passes through a glass prism. Similarly, sunlight produces a rainbow when it passes though raindrops, which act as miniature prisms.

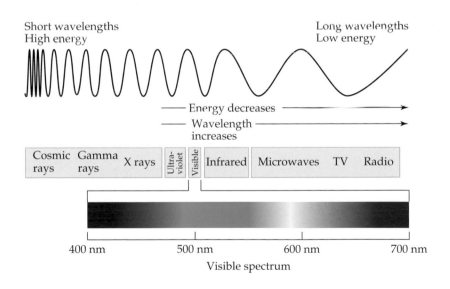

Figure 6.2 The Radiant Energy Spectrum The radiant energy spectrum includes short-wavelength cosmic rays through long-wavelength radio waves. The *ultraviolet* spectrum is approximately 100–400 nm, the visible spectrum is 400–700 nm, and the *infrared* region is 700–5000 nm.

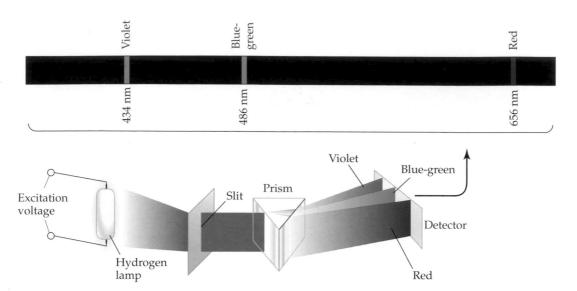

Figure 6.3 Hydrogen Line Spectrum An emission line spectrum is produced when hydrogen gas in a discharge tube is excited by electrical voltage. When the emitted light passes through a glass prism, we observe three discrete lines.

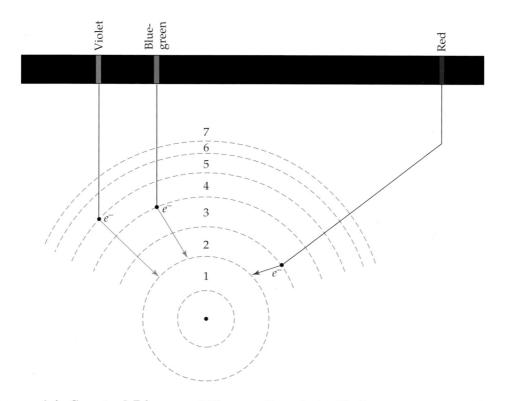

Figure 6.4 Spectral Lines and Energy Levels in Hydrogen When electrons drop from energy level **5** to **2**, we see a violet line. When electrons drop from level **4** to **2**, we see a blue-green line; from level **3** to **2**, we observe a red line.

Balmer Formula

In 1885, Johann Jakob Balmer, a Swiss mathematician and physicist, published a formula that accounts for the visible lines emitted from excited hydrogen gas. The **Balmer formula** shows that the wavelength (λ) of light for each line in the hydrogen spectrum is related to a small whole number (n) in the following way.

$$\frac{1}{\lambda} = \frac{1}{91 \text{ nm}} \left(\frac{1}{2^2} - \frac{1}{n^2} \right)$$

When Balmer set $n = 3$ in the formula, the calculated wavelength is that of the red line in the hydrogen spectrum. Similarly, substituting $n = 4$ and $n = 5$ gives values for the blue-green and violet lines. When Niels Bohr read Balmer's paper, he realized that the n value represents an energy level; and 2 in the formula corresponds to the second energy level.

The following example calculation illustrates the Balmer formula for an electron dropping from the sixth energy level to the second energy level in a hydrogen atom.

Example Exercise 6.1 • Balmer Formula

Calculate the wavelength of light corresponding to the energy released when an electron drops from $n = 6$ to $n = 2$ in a hydrogen atom.

Solution: We can use the Balmer formula to calculate the wavelength of the spectral line as follows.

$$\frac{1}{\lambda} = \frac{1}{91 \text{ nm}} \left(\frac{1}{2^2} - \frac{1}{n^2} \right)$$

$$\frac{1}{\lambda} = \frac{1}{91 \text{ nm}} \left(\frac{1}{2^2} - \frac{1}{6^2} \right)$$

$$\frac{1}{\lambda} = \frac{1}{91 \text{ nm}} \left(\frac{1}{4} - \frac{1}{36} \right)$$

$$\frac{1}{\lambda} = \frac{1}{91 \text{ nm}} (0.25 - 0.03)$$

$$\frac{1}{\lambda} = \frac{0.22}{91 \text{ nm}}$$

After taking the reciprocal and rounding to two significant digits, we obtain

$$\lambda = \frac{91 \text{ nm}}{0.22} = 413.6 = 410 \text{ nm}$$

A spectral line of this wavelength is in the violet portion of the spectrum. Although this line is not intense, if you look closely at the emission spectrum of hydrogen, you may see a faint violet line at 410 nm.

Rydberg Equation

In 1890, the Swedish physicist Johannes Rydberg derived a general formula for calculating the wavelengths of spectral lines from excited hydrogen gas. Rydberg was able to account for an electron dropping from any higher energy level (n_H) to any lower energy level (n_L). The Rydberg equation can be expressed as

$$\frac{1}{\lambda} = \frac{1}{91 \text{ nm}} \left(\frac{1}{n_L{}^2} - \frac{1}{n_H{}^2} \right)$$

The Balmer formula is adequate to explain the visible lines in the spectrum of hydrogen because excited electrons are dropping to $n = 2$. However, the Rydberg equation also accounts for electrons dropping to any energy level including $n = 1$. For example, when electrons drop from $n = 4$ to $n = 1$, the spectral line is in the ultraviolet region, which is not visible to the human eye. Similarly, when electrons drop from $n = 4$ to $n = 3$, the spectral line is in the infrared region of the spectrum, and is therefore not visible.

The following example calculation illustrates the Rydberg equation for an electron dropping from the fourth to the first energy level in a hydrogen atom.

Example Exercise 6.2 • Rydberg Equation

Calculate the wavelength of light released when an electron drops from $n = 4$ to $n = 1$ in a hydrogen atom.

Solution: We can use the Rydberg equation to find the wavelength of the line as follows.

$$\frac{1}{\lambda} = \frac{1}{91 \text{ nm}} \left(\frac{1}{n_L{}^2} - \frac{1}{n_H{}^2} \right)$$

$$\frac{1}{\lambda} = \frac{1}{91 \text{ nm}} \left(\frac{1}{1^2} - \frac{1}{4^2} \right)$$

$$\frac{1}{\lambda} = \frac{1}{91 \text{ nm}} \left(\frac{1}{1} - \frac{1}{16} \right)$$

$$\frac{1}{\lambda} = \frac{1}{91 \text{ nm}} \; (1 - 0.06)$$

$$\frac{1}{\lambda} = \frac{0.94}{91 \text{ nm}}$$

After taking the reciprocal and rounding to two significant digits, we obtain

$$\lambda = \frac{91 \text{ nm}}{0.94} = 96.8 = 97 \text{ nm}$$

A wavelength of 97 nm is below the visible spectrum (400–700 nm). If we refer to Figure 6.2, we find the line is in the *ultraviolet* region of the spectrum.

- Science Kit Hand Spectroscope
    The SK student spectroscope is available from Science Kit, Inc.
        @ 1-800-828-7777 (www.sciencekit.com).
- spectral tubes: hydrogen, helium, neon, argon, krypton, and mercury
- spectral tube power supply
    The spectrum tubes and power supply are readily available; for example,
    Sargent-Welch @ 1-800-SARGENT (www.sargentwelch.com), or
    VWR Scientific @ 1-800-932-5000 (www.vwr.com).
- (optional) colored pencils: violet, blue, green, yellow, orange, red

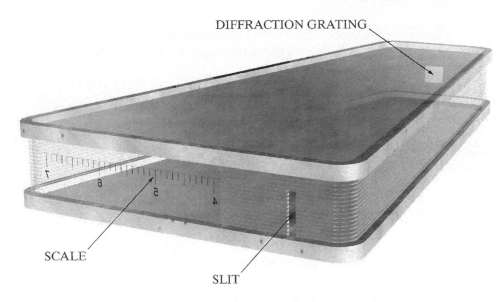

Figure 6.5 The Hand Spectroscope Light enters the spectroscope through the slit and strikes the diffraction grating. The grating is a thin piece of plastic with hundreds of parallel grooves that diffract light into different wavelengths.

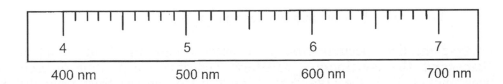

Figure 6.6 Wavelength Scale The spectroscope scale indicates wavelengths from 400 nm to 700 nm. The digit 4 is read as 400 nm, 5 as 500 nm, and 6 as 600 nm. There are ten subdivisions between each number on the scale; therefore each subdivision is 10 nm. For example, the scale divisions between 4 and 5 are read as 410, 420, 430, 440, 450, 460, 470, 480, and 490 nm.

A. Continuous Spectrum – White Light

1. With the hand spectroscope, observe the emission spectrum from one or more of the following: light from an overhead projector, light from an incandescent light bulb, and sunlight. Draw the observed spectrum on the wavelength scale in the Data Table.

   Note: It may be necessary to adjust the room lights in order to read the scale divisions in the hand spectroscope.

B. Line Spectrum – Hydrogen

1. Using a hand spectroscope, observe the hydrogen spectrum from a gas discharge tube. Draw the position of each line on the wavelength scale in the Data Table.

2. From the Balmer formula, find the wavelength (λ) of light produced when electrons drop from the 3rd to 2nd energy level. (Round the answer to two significant digits; for example, 655.2 nm rounds to 660 nm.)

3. Repeat the wavelength calculation for the spectral lines produced when electrons drop from the 4th to 2nd energy level; and from the 5th to 2nd energy level.

4. Record the observed and calculated wavelength values in the Data Table. State the error after comparing the observed and calculated wavelengths.

C. Line Spectra – Helium, Neon, Argon, Krypton, and Mercury

1. Insert a helium gas discharge tube into a spectral tube power supply. Using a hand spectroscope, observe the emission lines from helium gas. Draw the position of six intense lines on the wavelength scale in the Data Table.

2. Repeat the procedure using a neon gas discharge tube, argon gas discharge tube, krypton gas discharge tube, and mercury vapor discharge tube in the power supply. Draw the position of the most intense lines on the wavelength scale in the Data Table.

D. Identifying Unknown Elements in a Fluorescent Light

1. Observe the line spectrum from a fluorescent light using the hand spectroscope. **Disregard the continuous rainbow**, and draw the position of each line on the wavelength scale in the Data Table. Compare the line spectrum from the fluorescent light to the lines in the emission spectra of He, Ne, Ar, Kr, and Hg. Identify the elements in the fluorescent light from their "atomic fingerprints."

   Note: Currently, there are two elements in a fluorescent light, one of which should be easy to identify based on its "atomic fingerprint." The second element will require a careful comparison of "fingerprints" to identify.

PRELABORATORY ASSIGNMENT*

1. In your own words, define the following terms:

   "atomic fingerprint"

   Balmer formula

   continuous spectrum

   frequency

   light

   line spectrum

   photon

   visible spectrum

   wavelength

2. Arrange the six primary colors (**R, O, Y, G, B, V**) in the visible spectrum in order of

   (a) decreasing wavelength

   (b) decreasing frequency

   (c) decreasing energy

Answers in Appendix J

3. What wavelengths of light are indicated by the following scale readings observed through a hand spectroscope?

4. What are the colors of the three most intense lines in the emission spectrum of hydrogen?

5. A hydrogen gas discharge tube emits a faint line when electrons drop from the 10th to the 2nd energy level. Refer to Example Exercise 6.1 and calculate the wavelength for this spectral line.

6. How many photons are emitted when 1 electron drops from the 10th to 2nd energy level?

How many photons are emitted when 5 electrons drop from the 10th to 2nd energy level?

7. What safety precautions must be observed in this experiment?

NAME _____

DATE _____

SECTION _____

A. Continuous Spectrum – White Light   **Spectroscope #_____**

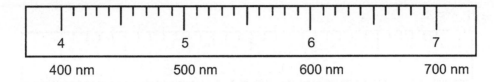

B. Line Spectrum – Hydrogen

   1. Observed Wavelengths of Spectral Lines

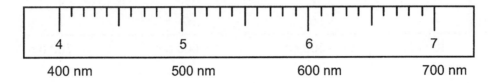

   2. Calculated Wavelengths of Spectral Lines (see Example Exercise 6.1)

     (a) red line (electrons drop from $n = 3$ to $n = 2$)

     (b) blue-green line (electrons drop from $n = 4$ to $n = 2$)

     (c) violet line (electrons drop from $n = 5$ to $n = 2$)

Spectral Line	Observed Wavelength	Calculated Wavelength	Experimental Error
red line	_____ nm	_____ nm	_____ nm
blue-green line	_____ nm	_____ nm	_____ nm
violet line	_____ nm	_____ nm	_____ nm

C. Line Spectra – Helium, Neon, Argon, Krypton, and Mercury

(a) Helium

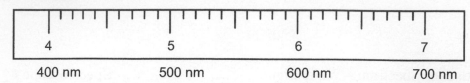

(b) Neon

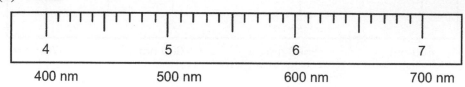

(c) Argon

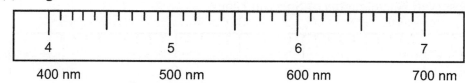

(d) Krypton

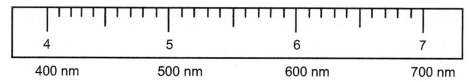

(e) Mercury

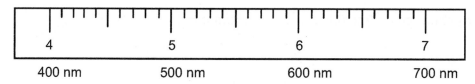

D. Identifying Unknown Elements in a Fluorescent Light

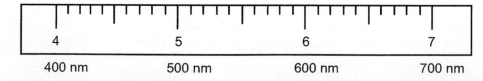

Based on its "atomic fingerprint," is **He** present in the fluorescent light?   YES / NO

Based on its "atomic fingerprint," is **Ne** present in the fluorescent light?   YES / NO

Based on its "atomic fingerprint," is **Ar** present in the fluorescent light?   YES / NO

Based on its "atomic fingerprint," is **Kr** present in the fluorescent light?   YES / NO

Based on its "atomic fingerprint," is **Hg** present in the fluorescent light?   YES / NO

NAME _____

SECTION _____

POSTLABORATORY ASSIGNMENT

1. In the spectrum from a hydrogen gas discharge tube (see Figure 6.4), what is the color of the emission line corresponding to the following changes in energy level:

   (a) electrons drop from level 3 to level 2?

   (b) electrons drop from level 4 to level 2?

   (c) electrons drop from level 5 to level 2?

2. Arrange each of the following in order of *increasing* energy:

   (a) the changes of electron energy levels from 5 to 2, 4 to 2, and 3 to 2.

   (least energetic) ____to____ > > > ____to____ > > > ____to____ *(most energetic)*

   (b) the wavelengths of spectral lines 650 nm, 480 nm, and 430 nm.

   (least energetic) _____ > > > _____ > > > _____ *(most energetic)*

3. Using the Rydberg equation (see Example Exercise 6.2), calculate the wavelength of the emission line when electrons drop from the 2nd to the 1st energy level in the hydrogen atom.

   Refer to Figure 6.2 and state whether the 2nd to 1st emission line is in the *ultraviolet, visible,* or *infrared* region of the spectrum.

4. Using the Rydberg equation (see Example Exercise 6.2), calculate the wavelength of the emission line when electrons drop from the 4th to the 3rd energy level in the hydrogen atom.

   Refer to Figure 6.2 and state whether the 4th to 3rd emission line is in the *ultraviolet, visible,* or *infrared* region of the spectrum.

5. Refer to the radiant energy spectrum in Figure 6.2 and state the region of the spectrum that is adjacent to ultraviolet, but is more energetic.

6. Refer to the radiant energy spectrum in Figure 6.2 and state the region of the spectrum that is adjacent to infrared, but is less energetic.

7. How many photons of light are emitted when the electrons in 1000 hydrogen atoms drop from energy level 5 to 2?

8. State the fundamental particle that represents each of the following:

   (a) the quantum nature of matter.

   (b) the quantum nature of electricity.

   (c) the quantum nature of light.

9. State whether the following instruments provides a continuous or quantized measurement.

   (a) metric ruler

   (b) platform balance

   (c) electronic balance

   (d) hypodermic syringe

   (e) digital clock

10. (optional) Advertising lights are often referred to as "neon lights." If advertising lights are gas discharge tubes, do all "neon lights" contain neon gas?   YES / NO

   Based on your observations of gas discharge tubes, suggest a gas suitable for:

   (a) a reddish-purple advertising light.

   (b) a reddish-orange advertising light.

   (c) a blue advertising light.

Families of Elements

- To study similar chemical properties for groups of elements in the periodic table.
- To observe flame tests and reactions for barium, calcium, lithium, potassium, sodium, and strontium solutions.
- To observe halide tests for bromide, chloride, and iodide solutions.
- To identify the alkali or alkaline earth element and the halide in an unknown solution.

DISCUSSION

In 1869, the Russian chemist Dmitri Mendeleev proposed that elements in the periodic table should be arranged by increasing atomic mass. In 1913, the English physicist Harry Moseley found that the elements should actually be arranged according to increasing atomic number. The so-called modern **periodic law** states that the properties of elements in the periodic table recur in a repeating pattern, when the elements are arranged according to *increasing atomic number.*

The elements in the periodic table are found in rows and columns. Elements in horizontal rows are called **periods**, or series. The elements in vertical columns are called **groups**, or families. Within each group of elements, or "family of elements," there are general trends in physical and chemical properties. For example, carbon, silicon, and germanium react with oxygen gas to give products with similar chemical formulas; that is, CO_2, SiO_2, and GeO_2. In this experiment, we will study three families of elements—**alkali metals**, **alkaline earth metals**, and **halides**. We will be able to identify those elements that belong to the same family by their chemical reactions.

In this experiment we will observe flame tests and solution reactions for selected alkali and alkaline earth elements. A **flame test** is performed by placing a small amount of solution on the end of a coiled wire. The wire is placed in a hot flame, and the color is observed (see Figure 7.1). For example, a solution containing a sodium compound gives a yellow flame test.

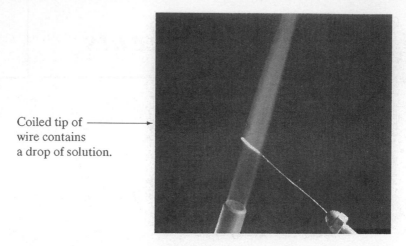

Coiled tip of
wire contains
a drop of solution.

Figure 7.1 Flame-Test Technique A drop of solution is placed on the coiled end of a wire, which is held at the tip of a hot burner flame. The color of the flame is characteristic for a given element; for example, sodium is yellow.

Although the colors are specific for each element, flame tests can be misleading. Sodium is often present as an impurity, so most flame tests will give a weak yellow flame. However, the intensity of the yellow flame for a sodium impurity is not as strong, and the distinction between a sodium impurity and an actual sodium sample can be made with little practice.

EQUIPMENT and CHEMICALS

- 13 × 100 mm test tubes (6) and test tube rack
- test tube brush
- flame-test wire (nichrome or platinum wire)
- wash bottle with distilled water

- ammonium carbonate solution 0.5 M $(NH_4)_2CO_3$
- ammonium phosphate solution 0.5 M $(NH_4)_2HPO_4$
- ammonium sulfate solution 0.5 M $(NH_4)_2SO_4$

- barium solution, 0.5 M $BaCl_2$
- calcium solution, 0.5 M $CaCl_2$
- lithium solution, 0.5 M $LiCl$
- potassium solution, 0.5 M KCl
- sodium solution, 0.5 M $NaCl$
- strontium solution, 0.5 M $SrCl_2$
- bromide solution, 0.5 M $NaBr$
- chloride solution, 0.5 M $NaCl$
- iodide solution, 0.1 M NaI
- hexane, C_6H_{14}
- dilute nitric acid, 6 M HNO_3
- chlorine water (bleach)
- unknown solutions

A. Analysis of Known Solutions

1. *Flame Tests of Known Solutions*

    (a) Place six test tubes in a test tube rack. Add 10 drops of barium, calcium, lithium, potassium, sodium, and strontium solutions into separate test tubes (Figure 7.2).

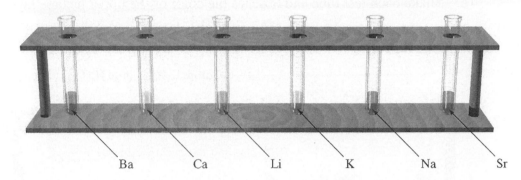

Ba   Ca   Li   K   Na   Sr

Figure 7.2 Alkali and Alkaline Earth Tests Solutions of barium, calcium, lithium, potassium, sodium, and strontium are placed in separate test tubes.

    (b) Obtain a flame-test wire and make a small loop in the end. Remove contamination by placing the wire loop at the tip of a burner flame. Continue to heat the wire until there is no longer any color produced in the flame.

    Note: If a flame-test wire continues to produce a colored flame, dip the wire into dilute hydrochloric acid and heat the wire to red heat.

    (c) Dip the clean wire into the test tube containing the barium solution. Place the loop at the tip of the flame. Record your observation. Clean the wire and repeat the flame test for calcium, lithium, potassium, sodium, and strontium solutions.

2. *Reactions of Known Solutions*

    (a) Add a few drops of ammonium carbonate, $(NH_4)_2CO_3$, solution in each test tube that was used for the flame test. If a precipitate forms, record *ppt* in the Data Table. If there is no reaction, record *NR*.

    (b) Clean the test tubes and rinse with distilled water. Put 10 drops of the barium, calcium, lithium, potassium, sodium, and strontium solutions into separate test tubes. Add a few drops of ammonium phosphate, $(NH_4)_3PO_4$, solution in each test tube. Record your observations in the Data Table.

    (c) Clean the test tubes and put 10 drops of barium, calcium, lithium, potassium, sodium, and strontium solutions in separate test tubes. Add a few drops of ammonium sulfate, $(NH_4)_2SO_4$, in each test tube and record your observations.

3. *Halide Tests of Known Solutions*

(a) Place three test tubes in a test tube rack. Add 10 drops of the following in separate test tubes: bromide solution, chloride solution, iodide solution.

(b) Add 10 drops of hexane, C_6H_{14}, 1 drop of nitric acid, HNO_3, and a few drops of chlorine water to each test tube (Figure 7.3).

(c) Shake each test tube and observe the color of the upper hexane layer.

Note: Dispose of the hexane layer in a waste container for organic chemicals.

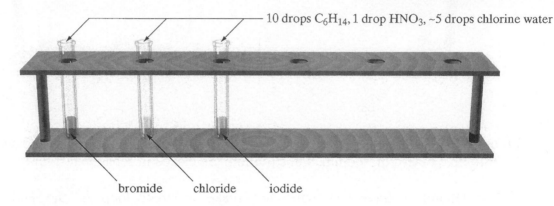

Figure caption labels: 10 drops C_6H_{14}, 1 drop HNO_3, ~5 drops chlorine water; bromide, chloride, iodide

Figure 7.3 Halide Tests Hexane, nitric acid, and chlorine water are added to bromide, chloride, and iodide solutions in separate test tubes.

B. Analysis of an Unknown Solution

1. *Flame Test*   Record the unknown number of a solution assigned by the Instructor. Perform a flame test on the solution, and record your observation in the Data Table.

2. *Solution Reactions*   Put 10 drops of unknown solution into three test tubes. Add a few drops of ammonium carbonate, $(NH_4)_2CO_3$, to the first test tube; a few drops of ammonium phosphate, $(NH_4)_3PO_4$, to the second; and a few drops of ammonium sulfate, $(NH_4)_2SO_4$, to the third. Record your observations in the Data Table.

3. *Halide Test*   Put 10 drops of unknown solution into a test tube. Add 10 drops of hexane, C_6H_{14}, 1 drop of nitric acid, HNO_3, and a few drops of chlorine water. Shake the test tube and record the color of the upper hexane layer.

EXPERIMENT 7 NAME ______________________

DATE ______________________ SECTION ______________________

*PRELABORATORY ASSIGNMENT**

1. In your own words, define the following terms:

   alkali metal

   alkaline earth metal

   flame test

   group

   halide

   immiscible

   period

   periodic law

   precipitate (ppt)

2. Which three alkali elements are investigated in this experiment?

   Which three alkaline earth elements?

   Which three halides?

3. Which element often causes difficulties in interpreting a flame test?

---

* *Answers in Appendix J*

4. Refer to **Figure 7.1** to answer the following.

   What color flame test indicates the presence of barium? _____

   What color flame test indicates the presence of calcium? _____

   What color flame test indicates the presence of lithium? _____

   What color flame test indicates the presence of potassium? _____

   What color flame test indicates the presence of sodium? _____

   What color flame test indicates the presence of strontium? _____

5. Are water and hexane miscible or immiscible?

   Is hexane present in the upper layer or lower layer?

   Do you observe the halide test in the upper layer or lower layer?

6. What safety precautions should be observed in this experiment?

NAME _____

SECTION _____

DATA TABLE

A. Analysis of Known Solutions

1. *Flame Tests of Known Solutions*

Solution Tested	Flame Test Observations
barium solution	_____
calcium solution	_____
lithium solution	_____
potassium solution	_____
sodium solution	_____
strontium solution	_____

2. *Reactions of Known Solutions*

Solution Tested	Solution Reaction Observations		
	ammonium carbonate	ammonium phosphate	ammonium sulfate
barium solution	_____	_____	_____
calcium solution	_____	_____	* _____
lithium solution	_____	_____	_____
potassium solution	_____	_____	_____
sodium solution	_____	_____	_____
strontium solution	_____	_____	_____

* Heat gently if NR.

3. *Halide Tests of Known Solutions*

Solution Tested	Hexane Layer Observations
bromide solution	_____
chloride solution	_____
iodide solution	_____

B. Analysis of an Unknown Solution   **UNKNOWN #** _____

1. *Flame Test of an Unknown Solution*

Solution Tested	Flame Test Observation
unknown solution	_____

2. *Reactions of an Unknown Solution*

	Solution Reaction Observations		
Solution Tested	ammonium carbonate	ammonium phosphate	ammonium sulfate
unknown solution	_____	_____	_____

Compare the flame test and solution reactions for the unknown to the observations in Procedures A.1 and A.2. Identify the alkali or the alkaline earth element present in the unknown solution; **circle** one of the following.

barium   calcium   lithium   potassium   sodium   strontium

3. *Halide Test of an Unknown Solution*

Solution Tested	Hexane Layer Observation
unknown solution	_____

Compare the hexane layer observation for the unknown solution to the observations in Procedure A.3. Identify the halide present in the unknown solution; **circle** one of the following.

bromide   chloride   iodide

POSTLABORATORY ASSIGNMENT

1. Refer to the reactions of alkali and alkaline earth elements in the data table (**Procedure A.2**). Circle the elements that produced chemical reactions similar to (a) barium and (b) lithium.

   (a)   barium   <u>Ca Li K Na Sr</u>   (b)   lithium   <u>Ba Ca K Na Sr</u>

2. What is the family name for the following groups of elements?

   (a)   Ba, Ca, Sr   _____   (b)   Li, K, Na   _____

3. What is the group number for the following families of elements?

   (a)   Ba, Ca, Sr   _____   (b)   Li, K, Na   _____

4. An unknown solution gave a strong yellow flame test. The solution gave no reaction with ammonium carbonate, ammonium phosphate, or ammonium sulfate. The halide test produced a yellow-orange color in the hexane layer. What is (a) the alkali or alkaline earth element, and (b) the halide in the unknown solution?

   (a)   _____   (b)   _____

5. An unknown solution gave a bright red flame test. The solution produced a white precipitate with ammonium carbonate, ammonium phosphate, and ammonium sulfate. The halide test produced a purple color in the hexane layer. What is (a) the alkali or alkaline earth element, and (b) the halide in the unknown solution?

   (a)   _____   (b)   _____

6. A fireworks display is produced by packing different chemicals and gunpowder in a rocket shell and exploding the mixture in the air. Based on your experimental observations, which elements could produce the following colors of fireworks?

   (a) brick-red   _____   (b) scarlet-red   _____

   (c) green   _____   (d) violet   _____

7. For the alkali metals, the metallic character (*increases / decreases*) up a group.
   For the alkaline earth metals, the atomic radius (*increases / decreases*) up a group.

8. For the Period 3 elements, the metallic character (*increases / decreases*) left to right.
   For the Period 4 elements, the atomic radius (*increases / decreases*) left to right.

9. Refer to the periodic table on the inside front cover of this lab manual. Select the symbol of the element that corresponds to the following description.

(a) the metal in Period 3, Group 13/IIIA   _____

(b) the semimetal in Period 3, Group 14/IVA   _____

(c) the nonmetal in Period 3, Group 15/VA   _____

(d) the nonmetal in Group 1/IA   _____

(e) the alkali metal in Period 3   _____

(f) the alkaline earth metal in Period 3   _____

(g) the halogen in Period 2   _____

(h) the noble gas in Period 2   _____

(i) the representative element in Period 4, Group 13/IIIA   _____

(j) the transition element in Period 4, Group 9/VIII   _____

(k) the rare earth element in Period 4   _____

(l) the lowest atomic mass lanthanide   _____

(m) the lowest atomic number actinide   _____

(n) the lowest atomic number transuranium element   _____

(o) the element with atomic number 82   _____

(p) the element with atomic mass 208.98 amu   _____

(q) the element with isotopic mass 209   _____

(r) the element filling a 3d sublevel in Group 6/VIB   _____

(s) the element with five valence electrons in Period 5   _____

(t) the element with ten 4d electrons in Period 5   _____

10. (optional) Refer to the periodic table on the inside front cover of this lab manual. Which two elements in the fifth period violate the original periodic law as stated by Mendeleev?

Identifying Cations in Solution

- To observe the chemical behavior of barium, calcium, and magnesium ions.
- To analyze an unknown solution for one or more of the following cations: Ba^{2+}, Ca^{2+}, and Mg^{2+}.
- To develop the following laboratory skills: centrifuging, flame testing, and using litmus paper.

Qualitative analysis is a systematic procedure for the separation and identification of ions present in an unknown solution. Cation analysis involves the separation and identification of each positively charged **cation** present in a sample.

If we have an **aqueous solution** containing different cations, it is possible to select a reagent that will form a **precipitate** with one of the cations, but not with the others. We can then use a **centrifuge** to separate the solid particles of precipitate from the aqueous solution. Thus, we separate the cation in the precipitate from the other cations in the original aqueous solution.

For example, we can separate the cations in a solution containing Ba^{2+}, Ca^{2+}, and Mg^{2+}, using ammonium sulfate. The sulfate ion, SO_4^{2-}, precipitates Ba^{2+}, but gives no reaction with either Ca^{2+} or Mg^{2+} cations (see Figure 8.1).

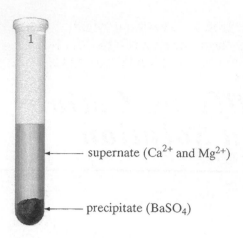

supernate (Ca^{2+} and Mg^{2+})

precipitate ($BaSO_4$)

Figure 8.1 Precipitation of $BaSO_4$ There is no reaction between Ca^{2+} and SO_4^{2-}, or between Mg^{2+} and SO_4^{2-}, because $CaSO_4$ and $MgSO_4$ are soluble.

When barium ions and sulfate ions are in a solution, a precipitate forms because barium sulfate, $BaSO_4$, is insoluble. If calcium and sulfate ions are in a solution, no precipitate forms because calcium sulfate, $CaSO_4$, is soluble; similarly, magnesium sulfate, $MgSO_4$, is soluble.

In this experiment, you will separate and identify Ba^{2+}, Ca^{2+}, and Mg^{2+}. First, a known solution containing all three cations will be analyzed to develop the necessary techniques. Second, an unknown solution with one or more of the three cations will be analyzed to determine the cations present.

Flame testing is a technique you can use to confirm the presence of an ion. A **flame test** is performed by dipping a wire into a solution and then holding the wire in a hot flame while observing the color produced (Figure 8.2). Many elements produce colored flames. For example, sodium is yellow, potassium is violet, and copper is green. Since sodium is always present as an impurity, flame tests are invariably contaminated by the yellow sodium flame.

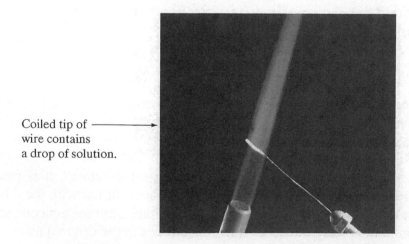

Coiled tip of wire contains a drop of solution.

Figure 8.2 Flame–Test Technique A drop of solution is placed on the coiled end of a wire, which is held at the tip of a hot burner flame. The color of the flame is characteristic for a given element; for example, sodium is yellow.

Litmus paper can be used to determine whether a solution is acidic or basic. A glass stirring rod is placed in the solution and touched to the litmus paper. Acidic solutions turn blue litmus paper red. Basic solutions turn red litmus paper blue (Figure 8.3).

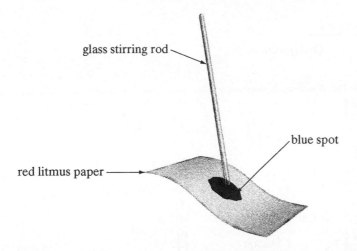

glass stirring rod

blue spot

red litmus paper

Figure 8.3 Litmus Paper Technique A glass stirring rod is placed in a solution and touched to red litmus paper. If the solution is basic, a blue spot is produced. If the solution is neutral or acidic, there is no change.

EQUIPMENT and CHEMICALS

- 13 × 100 mm test tubes (3) and test tube rack
- thin glass stirring rod
- wash bottle with distilled water
- centrifuge

- flame-test wire (nichrome or platinum wire)
- red litmus paper

- known cation solution (Ba^{2+}, Ca^{2+}, and Mg^{2+} as 0.1 M BaCl$_2$, CaCl$_2$, MgCl$_2$)

- ammonium sulfate solution, 0.1 M (NH$_4$)$_2$SO$_4$
- ammonium oxalate solution, 0.1 M (NH$_4$)$_2$C$_2$O$_4$
- sodium monohydrogen phosphate, 0.1 M Na$_2$HPO$_4$
- magnesium indicator (0.1 g *para*-nitrobenzeneazoresorcinol in 1 L of 0.025 M NaOH)
- dilute hydrochloric acid, 6 M HCl
- dilute sodium hydroxide, 6 M NaOH

- unknown cation solutions (Ba^{2+}, Ca^{2+}, and/or Mg^{2+} as 0.1 M BaCl$_2$, CaCl$_2$, MgCl$_2$)

We will begin the cation analysis with a known solution containing Ba^{2+}, Ca^{2+}, and Mg^{2+}. Ammonium sulfate is added to the known cation solution and the separation begins. Figure 8.4 presents an overview of the analysis. In **Step 1**, Ba^{2+} is confirmed; in **Step 2**, Ca^{2+} is confirmed; and in **Step 3**, Mg^{2+} is confirmed.

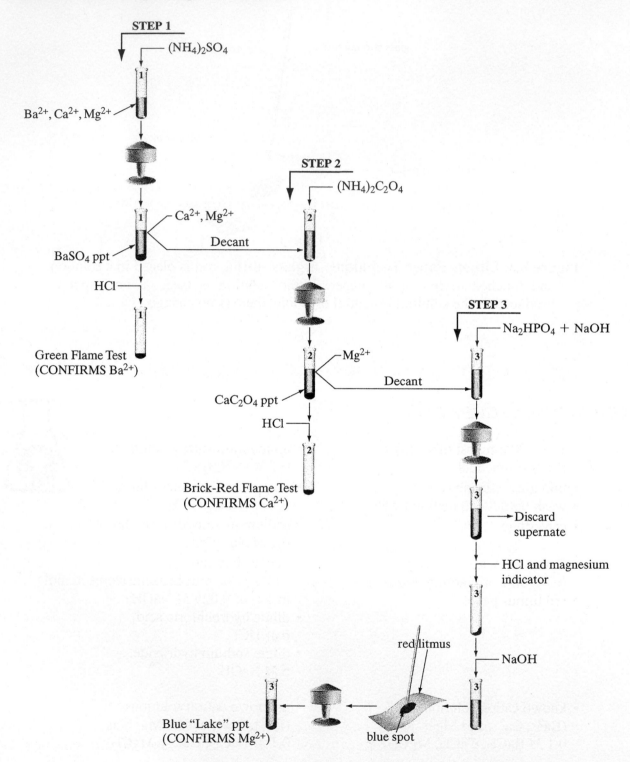

Figure 8.4 Cation Analysis The systematic separation and identification of Ba^{2+}, Ca^{2+}, and Mg^{2+} cations in a known solution.

General Directions: Clean three test tubes and a glass stirring rod with distilled water. Label the test tubes #1, #2, and #3. As a solution is analyzed, record the color of each precipitate in the Data Table.

A. Analysis of a Known Cation Solution

1. Identification of Ba^{2+} in a Known Solution

(a) Place 10 drops of the known solution in test tube #1. Add 10 drops of ammonium sulfate, $(NH_4)_2SO_4$, and mix with a glass stirring rod.

Note: A white precipitate suggests Ba^{2+} is present.

(b) Centrifuge, and then test for completeness of precipitation by adding another drop of ammonium sulfate. Pour off the supernate into test tube #2, and save for Step 2.

(c) Add 5 drops of dilute hydrochloric acid, HCl, to test tube #1, and stir thoroughly. Clean a flame-test wire with hydrochloric acid and dip it into the solution. Place the wire loop in a hot flame, and record the color.

Note: A green flame test confirms Ba^{2+} is present.

2. Identification of Ca^{2+} in a Known Solution

(a) Add 10 drops of ammonium oxalate, $(NH_4)_2C_2O_4$, to the solution in test tube #2.

Note: A white precipitate suggests Ca^{2+} is present.

(b) Centrifuge, and then test for completeness of precipitation by adding another drop of ammonium oxalate. Pour off the supernate into test tube #3, and save for Step 3.

(c) Add 5 drops of dilute hydrochloric acid to test tube #2, and stir thoroughly. Clean a flame-test wire with dilute HCl and dip the wire into the solution. Place the wire loop in a hot flame, and record the color.

Note: A brick-red flame test confirms Ca^{2+} is present.

3. Identification of Mg^{2+} in a Known Solution

(a) Add 10 drops of sodium monohydrogen phosphate, Na_2HPO_4, to the solution in test tube #3. Add 1 drop of sodium hydroxide, NaOH, and stir with a glass rod.

Note: A white precipitate suggests Mg^{2+} is present.

(b) Centrifuge, and discard the supernate.

(c) Dissolve the precipitate with dilute hydrochloric acid in test tube #3. Add a few drops of magnesium indicator. Add sodium hydroxide, NaOH, dropwise until the solution turns litmus paper blue. Centrifuge the precipitate.

Note: A blue gel precipitate confirms Mg^{2+} is present.

B. Analysis of an Unknown Cation Solution

1. *Identification of Ba^{2+} in an Unknown Solution*

    (a) Place 10 drops of unknown solution in test tube #1. Add 10 drops of ammonium sulfate, $(NH_4)_2SO_4$, and stir with a glass rod.

    Note: If there is no precipitate, Ba^{2+} is absent. Go directly to Step 2.

    (b) Centrifuge, and then test for completeness of precipitation by adding another drop of ammonium sulfate. Pour off the supernate into test tube #2, and save for Step 2.

    (c) Add 5 drops of dilute hydrochloric acid, HCl, to test tube #1 and stir thoroughly. Clean a flame-test wire with dilute HCl, and dip the wire into the solution. Place the wire loop in a hot flame, and record the color.

    Note: A green flame confirms Ba^{2+} is present.

2. *Identification of Ca^{2+} in an Unknown Solution*

    (a) Add 10 drops of ammonium oxalate, $(NH_4)_2C_2O_4$, to the solution in test tube #2.

    Note: If there is no precipitate, Ca^{2+} is absent. Go directly to Step 3.

    (b) Centrifuge, and then test for completeness of precipitation by adding another drop of ammonium oxalate. Pour off the supernate into test tube #3, and save for Step 3.

    (c) Add 5 drops of dilute hydrochloric acid, HCl, to test tube #2 and stir thoroughly. Clean a flame-test wire with dilute HCl, and dip the wire into the solution. Place the wire loop in a hot flame, and record the color.

    Note: A brick-red flame confirms Ca^{2+} is present.

3. *Identification of Mg^{2+} in an Unknown Solution*

    (a) Add 10 drops of sodium monohydrogen phosphate, Na_2HPO_4, to the solution in test tube #3. Add 1 drop of sodium hydroxide, NaOH, and stir with a glass rod.

    Note: If there is no precipitate, Mg^{2+} is absent.

    (b) Centrifuge, and discard the supernate.

    (c) Dissolve the precipitate with dilute hydrochloric acid in test tube #3. Add a few drops of magnesium indicator. Add sodium hydroxide, NaOH, dropwise until the solution turns litmus paper blue. Centrifuge the precipitate.

    Note: A blue "lake" precipitate confirms Mg^{2+} is present.

4. Based on the observations in steps 1–3, identify the cation(s) present in the unknown solution.

EXPERIMENT 8   NAME _____

DATE _____   SECTION _____

1. In your own words, define the following terms:

    aqueous solution

    cation

    centrifuge

    decant

    flame test

    precipitate (ppt)

    qualitative analysis

    supernate

2. Why is it necessary to use distilled water throughout this experiment?

3. Refer to **Figure 8.4** to answer the following.

    What color flame test confirms Ba^{2+} in test tube #1?   _____

    What color flame test confirms Ca^{2+} in test tube #2?   _____

    What color precipitate confirms Mg^{2+} in test tube #3?   _____

4. How is litmus paper used to test for a basic solution in test tube #3?

_____

Answers in Appendix J

5. Refer to **Figure 8.4**, and determine which of the following cations are present and absent in an unknown cation solution: Ba^{2+}, Ca^{2+}, and Mg^{2+}.

- Unknown solution in test tube #1 plus aqueous $(NH_4)_2SO_4$ gives a white precipitate.

- The supernate in test tube #1 is poured into test tube #2.

- The precipitate in test tube #1 plus HCl gives a green flame test for two seconds.

- The solution in test tube #2 plus aqueous $(NH_4)_2C_2O_4$ gives no reaction.

- The solution in test tube #2 is poured into test tube #3.

- Test tube #3 plus aqueous Na_2HPO_4 and NaOH gives a white precipitate.

- The precipitate dissolves in HCl, and magnesium indicator is added.

- The solution is made basic with aqueous NaOH and centrifuged.

- A blue "lake," that is a clear blue gel, is observed at the bottom of the test tube.

Cation(s) **present** _____   Cation(s) **absent** _____

6. What safety precautions should be taken while performing this experiment?

NAME _____

SECTION _____

DATA TABLE

A. Analysis of a Known Cation Solution

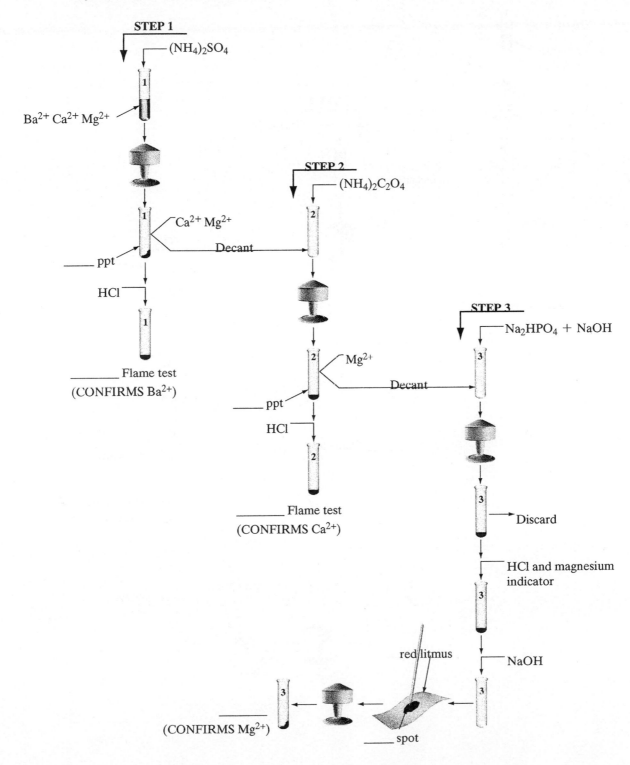

B. Analysis of an Unknown Cation Solution

UNKNOWN # _____

Cation(s) **present** _____

Cation(s) **absent** _____

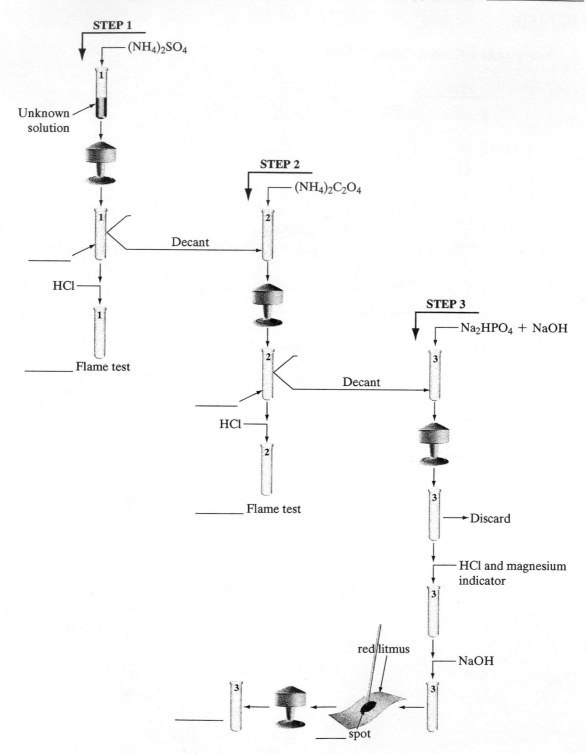

POSTLABORATORY ASSIGNMENT

1. A "barium/calcium/magnesium" unknown solution and aqueous $(NH_4)_2SO_4$ produced a white precipitate that gave a green flame test. After centrifuging and decanting, the addition of aqueous $(NH_4)_2C_2O_4$ produced a white precipitate that gave a brick-red flame test. After centrifuging and decanting, the addition of aqueous Na_2HPO_4 and $NaOH$ gave no reaction. State the cation(s) confirmed present and absent in the unknown solution.

   Cation(s) **present** _____   Cation(s) **absent** _____

2. A "barium/calcium/magnesium" unknown solution gave no reaction with aqueous $(NH_4)_2SO_4$, or aqueous $(NH_4)_2C_2O_4$. Adding aqueous Na_2HPO_4 and $NaOH$ gave a white precipitate. After dissolving the precipitate in HCl, and adding magnesium reagent, the solution gave a clear blue gel precipitate. State the cation(s) confirmed present and absent in the unknown solution.

   Cation(s) **present** _____   Cation(s) **absent** _____

3. Supply the Stock systematic name for the following cations.

   (a) Co^{3+} _____   (b) Cu^{2+} _____

   (c) Fe^{2+} _____   (d) Pb^{4+} _____

   (e) Hg_2^{2+} _____   (f) Sn^{2+} _____

   (g) Ni^{2+} _____   (h) Mn^{2+} _____

   (i) Cr^{3+} _____   (j) Ag^+ _____

4. Provide the formula for each of the following cations.

   (a) cuprous ion _____   (b) cupric ion _____

   (c) ferrous ion _____   (d) ferric ion _____

   (e) plumbous ion _____   (f) plumbic ion _____

   (g) mercurous ion _____   (h) mercuric ion _____

   (i) stannous ion _____   (j) stannic ion _____

5. Complete the table below as shown by the example. Combine the ions into a correct formula, and name the compound.

	NO_3^-	CO_3^{2-}	PO_4^{3-}
Ba^{2+}	$Ba(NO_3)_2$ *barium nitrate*		
Ca^{2+}			
Mg^{2+}			

6. (optional) Complete the table below as shown by the example. Combine the ions into a correct formula, and name the compound.

	chromate ion	permanganate ion	cyanide ion
mercury(I) ion	Hg_2CrO_4 *mercury(I) chromate*		
cobalt(II) ion			
iron(III) ion			

Identifying Anions in Solution

OBJECTIVES

- To observe the chemical behavior of iodide, chloride, and sulfate ions.
- To analyze an unknown solution for one or more of the following anions:
  I^-, Cl^-, and SO_4^{2-}.
- To develop the following laboratory skills: centrifuging, washing a precipitate, and testing with litmus paper.

DISCUSSION

Qualitative analysis is a systematic procedure for the separation and identification of ions present in an unknown solution. Anion analysis involves the separation and identification of each negatively charged **anion** present in a sample.

If we have an **aqueous solution** containing different anions, it is possible to select a reagent that will form a **precipitate** with one of the anions, but not with the others. We can then use a **centrifuge** to separate the solid particles of precipitate from the aqueous solution. Thus, we separate the anion in the precipitate from the other anions in the original aqueous solution.

For example, we can separate the anions in a solution containing I^-, Cl^-, and SO_4^{2-}, using silver nitrate. The silver ion, Ag^+, precipitates I^- and Cl^-, but gives no reaction with the SO_4^{2-} anion (see Figure 9.1).

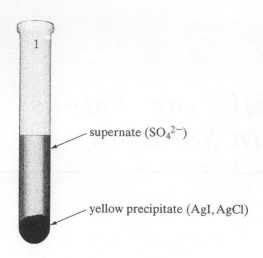

Figure 9.1 Precipitation of AgI and AgCl There is no reaction between Ag^+ and SO_4^{2-} because Ag_2SO_4 is soluble.

When silver ions and iodide ions are in a solution, a precipitate forms because silver iodide, AgI, is insoluble. Similarly, silver ions and chloride ions in a solution give a precipitate of silver chloride, AgCl. If silver ions and sulfate ions are in a solution, no precipitate forms because silver sulfate, Ag_2SO_4, is soluble.

In this experiment, you will separate and identify I^-, Cl^-, and SO_4^{2-}. First, a known solution containing all three anions will be analyzed to develop the necessary techniques. Second, an unknown solution with one or more of the three anions will be analyzed to determine the anions present.

Litmus paper can be used to determine whether a solution is acidic or basic. A glass stirring rod is placed in the solution and touched to the litmus paper. Acidic solutions turn blue litmus paper red. Basic solutions turn red litmus paper blue (Figure 9.2).

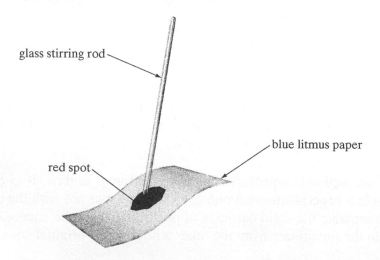

Figure 9.2 Litmus Paper Technique A glass stirring rod is placed in a solution and touched to blue litmus paper. If the solution is acidic, a red spot is produced. If the solution is neutral or basic, there is no change.

We will begin the anion analysis with a known solution containing I^-, Cl^-, and SO_4^{2-}. First, we add aqueous silver nitrate to the known anion solution and the separation begins. Figure 9.3 presents an overview of the analysis. In **Step 1**, I^- is confirmed; in **Step 2**, Cl^- is confirmed; and in **Step 3**, SO_4^{2-} is confirmed.

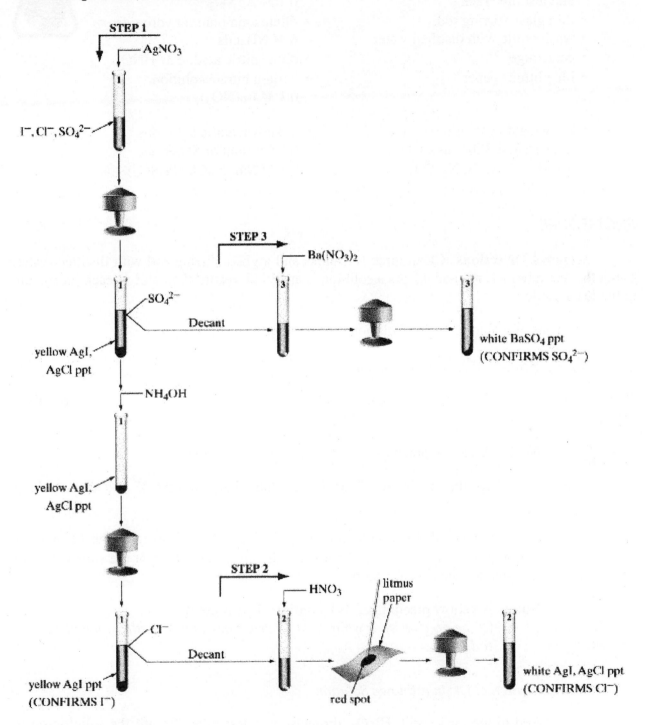

Figure 9.3 Anion Analysis The systematic separation and identification of I^-, Cl^-, and SO_4^{2-} anions in a known solution.

- 13 × 100 mm test tubes (3) and test tube rack
- thin glass stirring rod
- wash bottle with distilled water
- centrifuge
- blue litmus paper

- silver nitrate solution, 0.1 M $AgNO_3$
- dilute ammonium hydroxide, 6 M NH_4OH
- dilute nitric acid, 6 M HNO_3
- barium nitrate solution, 0.1 M $Ba(NO_3)_2$

- known anion solution (I^-, Cl^-, and SO_4^{2-} as 0.1 M NaI, NaCl, Na_2SO_4)

- unknown anion solutions (I^-, Cl^-, and/or SO_4^{2-} as 0.1 M NaI, NaCl, Na_2SO_4)

PROCEDURE

General Directions: Clean three test tubes and a glass stirring rod with distilled water. Label the test tubes #1, #2, and #3. As a solution is analyzed, record the color of each precipitate in the Data Table.

A. Analysis of a Known Anion Solution

1. *Identification of I⁻ in a Known Solution*

   (a) Place 10 drops of the known solution in test tube #1. Add 20 drops of silver nitrate, $AgNO_3$, and mix with a glass stirring rod.

   Note: A yellow precipitate, AgI, suggests I⁻ is present.

   (b) Centrifuge the precipitate. Pour the supernate into test tube #3 and save for Step 3.

   (c) Add 10 drops of dilute ammonium hydroxide, NH_4OH, to test tube #1 and stir thoroughly with a glass rod. Centrifuge the precipitate. Decant the supernate into test tube #2 and save for Step 2.

   Note: A yellow precipitate, AgI, confirms I⁻ is present.
   (*If the precipitate is white, add 10 drops of water and stir with a glass rod.*)

2. *Identification of Cl⁻ in a Known Solution*

   Add dilute nitric acid, HNO_3, dropwise into test tube #2 until the solution turns litmus paper red. Centrifuge the precipitate.

   Note: A white precipitate, AgCl, confirms Cl⁻ is present.
   (*If the precipitate is yellow, it contains AgI particles from test tube #1.*)

3. Identification of SO_4^{2-} in a Known Solution

Add 10 drops of barium nitrate, $Ba(NO_3)_2$, to the solution in test tube #3. Centrifuge the precipitate.

Note: A white precipitate, $BaSO_4$, confirms SO_4^{2-} is present.
(*If the precipitate is yellow, it contains AgI particles from test tube #1.*)

B. Analysis of an Unknown Anion Solution

1. Identification of I^- in an Unknown Solution

(a) Place 10 drops of unknown solution in test tube #1. Add 20 drops of silver nitrate, $AgNO_3$, and stir with a glass rod.

Note: If there is no precipitate, I^- and Cl^- are absent. Go directly to Step 3.

(b) Centrifuge the precipitate. Pour the supernate into test tube #3, and save for Step 3.

(c) Add 10 drops of dilute ammonium hydroxide, NH_4OH, to test tube #1, and stir thoroughly with a glass rod. Centrifuge the precipitate. Decant the supernate into test tube #2, and save for Step 2.

Note: If there is no precipitate, I^- is absent. Go directly to Step 2.

2. Identification of Cl^- in an Unknown Solution

Add dilute nitric acid, HNO_3, dropwise to test tube #2 until the solution turns litmus paper red. Centrifuge the precipitate.

Note: If there is no precipitate, Cl^- is absent. Go directly to Step 3.

3. Identification of SO_4^{2-} in an Unknown Solution

Add 10 drops of barium nitrate, $Ba(NO_3)_2$, to the solution in test tube #3. Centrifuge the precipitate.

Note: If there is no precipitate, SO_4^{2-} is absent.

4. Based on the observations in steps 1–3, identify the anion(s) present in the unknown solution.

EXPERIMENT 9 NAME ______________________

DATE ______________ SECTION ______________________

*PRELABORATORY ASSIGNMENT**

1. In your own words, define the following terms:

   anion

   aqueous solution

   centrifuge

   decant

   precipitate (ppt)

   qualitative analysis

   supernate

2. Why is it necessary to use distilled water throughout the experiment?

3. Refer to **Figure 9.3** to answer the following.

   What color precipitate confirms I^- in test tube #1? ______________

   What color precipitate confirms Cl^- in test tube #2? ______________

   What color precipitate confirms SO_4^{2-} in test tube #3? ______________

4. How is litmus paper used to test for an acidic solution in test tube #2?

---

* *Answers in Appendix J*

5. Refer to **Figure 9.3**, and determine which of the following anions are present and absent in an unknown anion solution: I^-, Cl^-, and SO_4^{2-}.

- Unknown solution in test tube #1 plus aqueous $AgNO_3$ gives a yellow precipitate.

- The supernate in test tube #1 is poured into test tube #3.

- The yellow precipitate in test tube #1 does not dissolve in aqueous NH_4OH.

- The supernate in test tube #1 is decanted into test tube #2.

- Test tube #2 plus aqueous HNO_3 gives no reaction.

- Test tube #3 plus aqueous $Ba(NO_3)_2$ yields a white precipitate.

Anion(s) **present** _____   Anion(s) **absent** _____

6. What safety precautions should be taken while performing this experiment?

DATA TABLE

A. Analysis of a Known Anion Solution

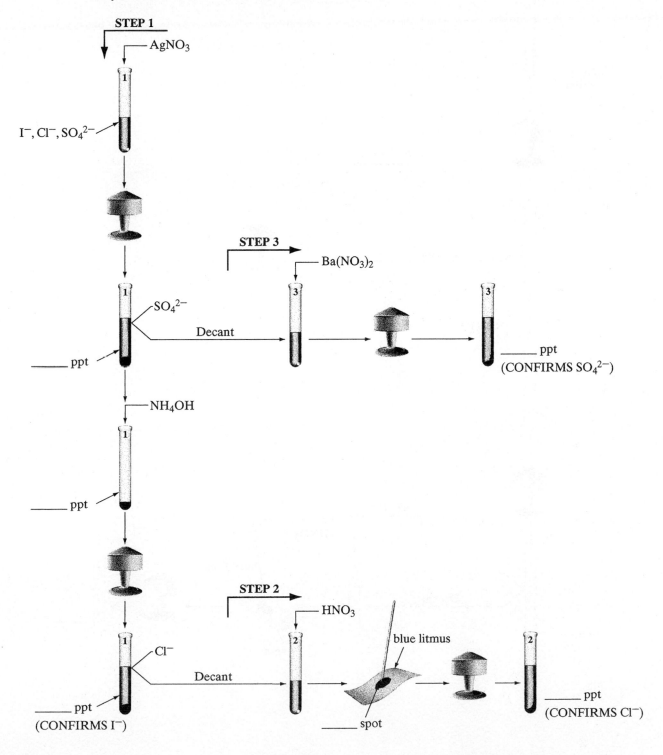

STEP 1

AgNO$_3$

I^-, Cl^-, SO_4^{2-}

STEP 3

Ba(NO$_3$)$_2$

SO$_4^{2-}$

Decant

_____ ppt

_____ ppt
(CONFIRMS SO$_4^{2-}$)

NH$_4$OH

_____ ppt

STEP 2

HNO$_3$

Cl$^-$

Decant

blue litmus

_____ spot

_____ ppt
(CONFIRMS Cl$^-$)

_____ ppt
(CONFIRMS I$^-$)

B. Analysis of an Unknown Anion Solution

Anion(s) **present** _____   Anion(s) **absent** _____

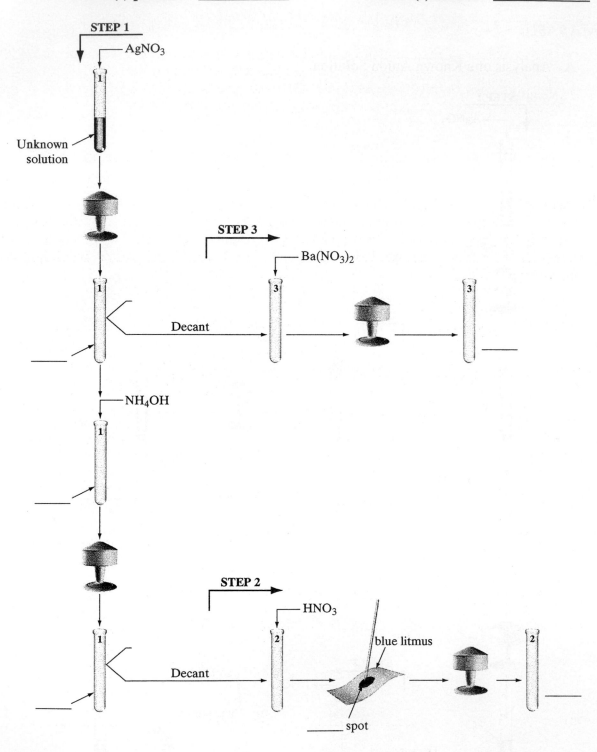

NAME _____

SECTION _____

POSTLABORATORY ASSIGNMENT

1. An "iodide/chloride/sulfate" unknown solution gave a yellow precipitate with aqueous $AgNO_3$, and after centrifuging, the supernate gave no reaction with aqueous $Ba(NO_3)_2$. The yellow precipitate did not dissolve in aqueous NH_4OH, and adding aqueous HNO_3 gave no reaction. State the anion(s) confirmed present and absent in the unknown solution.

   Anion(s) **present** _____   Anion(s) **absent** _____

2. An "iodide/chloride/sulfate" unknown solution gave a white precipitate with aqueous $AgNO_3$, and after centrifuging, the supernate gave a white precipitate with aqueous $Ba(NO_3)_2$. The original white precipitate dissolved completely in aqueous NH_4OH, and produced a white precipitate after adding aqueous HNO_3. State the anion(s) confirmed present and absent in the unknown solution.

   Anion(s) **present** _____   Anion(s) **absent** _____

3. Supply a systematic name for the following common monoatomic anions.

   (a) F^- _____   (b) Cl^- _____

   (c) Br^- _____   (d) I^- _____

   (e) O^{2-} _____   (f) S^{2-} _____

   (g) N^{3-} _____   (h) P^{3-} _____

4. Provide the formula for the following common polyatomic anions.

   (a) nitrate ion   _____   (b) nitrite ion   _____

   (c) sulfate ion   _____   (d) sulfite ion   _____

   (e) chlorate ion   _____   (f) chlorite ion   _____

   (g) perchlorate ion   _____   (h) hypochlorite ion _____

5. Complete the table below as shown by the example. Combine the ions into a correct formula, and name the compound.

	I^-	Cl^-	SO_4^{2-}
Li^+	LiI *lithium iodide*		
Sr^{2+}			
Al^{3+}			

6. (optional) Complete the table below as shown by the example. Combine the ions into a correct formula, and name the compound.

	acetate ion	dichromate ion	hydroxide ion
ammonium ion	$NH_4C_2H_3O_2$ *ammonium acetate*		
cadmium ion			
lead(IV) ion			

Analysis of a Penny

- To become familiar with the evidence for chemical reaction.
- To translate word equations into balanced chemical equations.
- To determine the percentages of copper and zinc in a "zinc penny."
- To gain experience in observing evidence for chemical reactions.

Most ordinary chemical reactions can be classified as one of five basic types. The first type of reaction occurs when two or more **reactants** combine to form a single **product**. This type of reaction is called a *combination reaction*.

$$A + Z \rightarrow AZ$$

A second type of reaction occurs when a single compound breaks down into two or more simpler substances, often with the use of a **catalyst** to speed up the reaction. This type is called a *decomposition reaction*.

$$AZ \rightarrow A + Z$$

A third type of reaction occurs when one element displaces another. For this to occur, a more active element that is higher in the **activity series** displaces an element that is lower in the series. This type is called a *single-replacement reaction*.

$$A + BZ \rightarrow AZ + B$$

A fourth type of reaction occurs when two substances in **aqueous solution** switch partners; that is, an anion of one substance exchanges with another. Usually one of the products is an insoluble **precipitate**. This type is called a *double-replacement reaction.*

$$AX + BZ \rightarrow AZ + BX$$

A fifth type of reaction occurs when an acid and a base react to form a salt and water. This is a special type of double-replacement reaction, and is called a *neutralization reaction.*

$$HX + BOH \rightarrow BX + HOH$$

Notice that the hydrogen ion in the acid neutralizes the hydroxide ion in the base to form water. If water is written as HOH, the neutralization is obvious and the equation may be easier to balance.

In this experiment, you will carefully observe and record evidence for a chemical reaction. Evidence for a reaction may include any of the following: (1) a gas is produced; (2) a precipitate is formed; (3) a color change is observed; (4) an energy change is noted. In order to describe the reaction, we use various symbols in the chemical equation. Table 10.1 lists some of these.

Table 10.1 Symbols in Chemical Equations

Symbol	Translation
$\rightarrow$	produces, yields (the arrow separates reactants from products)
+	added to, reacts with (separates two or more reactants or products)
Δ	heat (written above $\rightarrow$)
NR	no reaction (written after $\rightarrow$)
(s)	solid or precipitate
(l)	liquid
(g)	gas
(aq)	aqueous solution

In order to write a chemical equation, it is necessary to predict the products from a reaction. Initially, this may be a difficult task. To aid you in writing chemical equations, the products are given for each reaction. You only need to translate the given word equations into balanced chemical equations. The following examples will illustrate.

Combination Reaction

$$\text{calcium(s)} + \text{oxygen(g)} \rightarrow \text{calcium oxide(s)}$$
$$2\,Ca(s) + O_2(g) \rightarrow 2\,CaO(s)$$

Decomposition Reaction

$$\text{lithium hydrogen carbonate(s)} \rightarrow \text{lithium carbonate(s)} + \text{water(g)} + \text{carbon dioxide(g)}$$
$$2\,LiHCO_3(s) \rightarrow Li_2CO_3(s) + H_2O(g) + CO_2(g)$$

Single-Replacement Reaction

$$\text{tin(s)} + \text{hydrochloric acid(aq)} \rightarrow \text{tin(II) chloride(aq)} + \text{hydrogen(g)}$$
$$Sn(s) + 2\,HCl(aq) \rightarrow SnCl_2(aq) + H_2(g)$$

potassium carbonate(aq) + calcium chloride(aq) $\rightarrow$ calcium carbonate(s) + potassium chloride(aq)

$$K_2CO_3(aq) + CaCl_2(aq) \rightarrow CaCO_3(s) + 2\,KCl(aq)$$

Neutralization Reaction

nitric acid(aq) + barium hydroxide(aq) $\rightarrow$ barium nitrate(aq) + water

$$2\,HNO_3(aq) + Ba(OH)_2(aq) \rightarrow Ba(NO_3)_2(aq) + 2\,HOH(l)$$

Analysis of a "Zinc Penny"

In 1982 the United States Mint stopped making copper pennies because of the high price of copper, and began phasing in pennies made of zinc. The U.S. Mint cast the penny from pure zinc and electroplated the zinc with a layer of copper only 0.01 mm thick! The resulting "zinc penny" has the same appearance as a copper penny, however, it is about 20% lighter in mass.

In this experiment you will cut a "zinc penny" to expose the zinc and drop the penny into sulfuric acid. Although copper does not react with acid, zinc does react with sulfuric acid and leaves a thin copper shell. The chemical equations for the two reactions are:

$$Cu(s) + H_2SO_4(aq) \rightarrow NR$$
$$Zn(s) + H_2SO_4(aq) \rightarrow ZnSO_4(aq) + H_2(g)$$

The following example exercise illustrates the calculation for the percentages of copper and zinc in a "zinc penny."

Example Exercise 10.1 • Percent Composition of a "Zinc Penny"

A 1995 penny having a mass of 2.536 g is cut to expose the zinc and dropped into sulfuric acid. After the zinc has reacted, the copper shell is found to have a mass of 0.063 g. Calculate the percentages of copper and zinc in the "zinc penny."

Solution: The percentage of copper is simply the ratio of the mass of Cu metal to the mass of the penny; that is,

$$\frac{0.063\ \cancel{g}}{2.536\ \cancel{g}} \times 100\% = 2.5\%\ Cu$$

The percentage of zinc is the ratio of the mass of Zn metal to the mass of the original penny. The mass of Zn corresponds to the mass loss of the penny: 2.536 g – 0.063 g = 2.473 g. Thus,

$$\frac{2.473\ \cancel{g}}{2.536\ \cancel{g}} \times 100\% = 97.52\%\ Zn$$

Experimentally, the 1995 "zinc penny" is 2.5% Cu and 97.52% zinc.

Students often ask if it is illegal to destroy a coin. According to a U.S. Treasury official: *"the law provides criminal penalties for anyone who **fraudulently** alters, defaces, mutilates, impairs, diminishes, falsifies, scales or lightens any of the coins coined at the mints of the United States."* Since we are not intending to defraud, this experiment is **legal**.

A. Instructor Demonstrations

- crucible tongs
- deflagrating spoon

- magnesium, Mg ribbon
- sulfur, S powder

B–F. Student Experiments

- 13 × 100 mm test tubes (6)
- test tube holder
- test tube brush
- wash bottle with distilled water
- 250-mL Erlenmeyer flask
- 100-mL beaker
- copper(II) sulfate pentahydrate, solid $CuSO_4 \cdot 5H_2O$
- sodium hydrogen carbonate, solid $NaHCO_3$
- wooden splints
- copper, Cu wire
- magnesium, Mg turnings
- calcium, Ca turnings

- hydrochloric acid, 0.1 M HCl
- silver nitrate, 0.1 M $AgNO_3$
- copper(II) nitrate, 0.1 M $Cu(NO_3)_2$
- aluminum nitrate, 0.1 M $Al(NO_3)_3$
- potassium carbonate, 0.5 M K_2CO_3
- sodium phosphate, 0.5 M Na_3PO_4
- nitric acid, 0.1 M HNO_3
- sulfuric acid, 0.1 M H_2SO_4
- phosphoric acid, 0.1 M H_3PO_4
- sodium hydroxide, 0.5 M NaOH
- phenolphthalein acid-base indicator
- "zinc penny" (post-1982 mint date)
- dilute sulfuric acid, 3 M H_2SO_4
- acetone, C_3H_6O

General Directions: For Procedures A–E, record your observations in the Data Table. Since the "zinc penny" may require three hours to react completely, it is advisable to first perform Procedure F, and then begin Procedures A–E.

A. Combination Reactions—Instructor Demonstrations

1. Hold a 2-cm strip of magnesium ribbon with crucible tongs, and ignite the metal in a hot burner flame.
2. Put about 1 g of sulfur in a deflagrating spoon. Dim the lights and ignite the powder with a hot burner flame. Place the burning sulfur under a fume hood to avoid the pungent odor of the sulfur dioxide gas.

> **Note:** The Instructor should demonstrate or closely supervise each of these exothermic reactions.

B. Decomposition Reactions

1. Put a pea-sized portion of copper(II) sulfate pentahydrate crystals into a dry test tube. Grasp the test tube with a test tube holder and heat with a burner (see Figure 1.1). Note the color change, and observe the inside surface at the top of the test tube.
2. Add sodium hydrogen carbonate (baking soda) into a 250-mL Erlenmeyer flask so as to cover the bottom of the flask. Support the flask on a ring stand, using a wire gauze.
    (a) Hold a flaming splint in the mouth of the flask, and record how long it burns.
    (b) Heat the flask strongly with the laboratory burner until moisture is observed; hold a flaming splint in the mouth of the flask, and record how long it burns.

C. Single-Replacement Reactions

1. Put 20 drops of silver nitrate solution into a test tube, and add a small piece of copper wire. Allow a few minutes for reaction and then record your observation.
2. Put 20 drops of hydrochloric acid into a test tube, and add a small piece of magnesium metal. Record your observation.
3. Put 20 drops of distilled water into a test tube, and add a small piece of calcium metal. Record your observation.

D. Double-Replacement Reactions

1–3. Put 10 drops of silver nitrate, copper(II) nitrate, and aluminum nitrate solutions into separate test tubes #1–3. Add a few drops of potassium carbonate solution to test tubes #1, #2, and #3. Observe the reactions, and record your observations.
4–6. Put 10 drops of silver nitrate, copper(II) nitrate, and aluminum nitrate solutions into separate test tubes #4–6. Add a few drops of sodium phosphate solution to test tubes #4, #5, and #6. Observe the reactions, and record your observations.

E. Neutralization Reactions

1. Put 10 drops of nitric acid, sulfuric acid, and phosphoric acid into separate test tubes #1–3. Add one drop of phenolphthalein to each of the test tubes. Add drops of dilute sodium hydroxide solution into test tube #1 until a permanent color change is observed.

    Note: Phenolphthalein is an acid–base indicator that is colorless in acidic and neutral solutions, and pink in basic solutions.

2. Add drops of dilute sodium hydroxide solution to test tube #2 until a permanent color change is observed.
3. Add drops of dilute sodium hydroxide solution to test tube #3 until a permanent color change is observed.

F. Percentages of Copper and Zinc in a Penny

1. Obtain a post-1982 penny, and record the mint date. Use metal shears to cut the coin as shown in Figure 10.1.
2. Accurately weigh the penny on a balance, and record the mass.

Figure 10.1 Exposing Zinc in a Penny A "zinc penny" should be cut as shown to ensure a rapid and complete reaction with sulfuric acid.

3. Drop the penny into a 100-mL beaker, and add about 20 mL of dilute sulfuric acid. The reaction requires about 3 hours for the zinc in the coin to react completely.

   Caution: If acid contacts your skin, wash immediately with water.

4. When the coin stops producing gas bubbles, discard the sulfuric acid in the sink.
5. Wash the coin with distilled water, and discard the wash solution. Rinse the coin with acetone, and discard the rinse solution. When the coin appears dry, weigh the copper shell and record the mass in the Data Table.
6. Calculate the percentages of copper and zinc in the penny.

NAME _____

SECTION _____

1. In your own words, define the following terms:

    activity series

    catalyst

    precipitate (ppt)

    product

    reactant

2. Explain the meaning of the following symbols:

    $\rightarrow$

    $+$

    Δ

    NR

    (g)

    (l)

    (s)

    (aq)

3. List four observations that are evidence of a chemical reaction.

    (a)

    (b)

    (c)

    (d)

_____

* Answers in Appendix J

4. What color is phenolphthalein acid-base indicator in:

   (a) an acidic solution?

   (b) a basic solution?

5. What is the acceptable range of mint dates for a "zinc penny"?

6. A 1990 penny has a mass of 2.545 g and produces a copper shell with a mass of 0.064 g. Refer to Example Exercise 10.1 and calculate the percentage of (a) copper metal and (b) zinc metal in the penny.

7. What safety precautions must be observed in this experiment?

NAME _____

SECTION _____

DATA TABLE

Procedure	Observation

A. Combination Reactions—Instructor Demonstrations

   1. $Mg + O_2 \xrightarrow{\Delta}$ _____

   2. $S + O_2 \xrightarrow{\Delta}$ _____

B. Decomposition Reactions

   1. $CuSO_4 \cdot 5H_2O \xrightarrow{\Delta}$ _____

   2. (a) $NaHCO_3$ _____

      (b) $NaHCO_3 \xrightarrow{\Delta}$ _____

C. Single-Replacement Reactions

   1. $Cu + AgNO_3 \rightarrow$ _____

   2. $Mg + HCl \rightarrow$ _____

   3. $Ca + H_2O \rightarrow$ _____

D. Double-Replacement Reactions

   1. $AgNO_3 + K_2CO_3 \rightarrow$ _____

   2. $Cu(NO_3)_2 + K_2CO_3 \rightarrow$ _____

   3. $Al(NO_3)_3 + K_2CO_3 \rightarrow$ _____

   4. $AgNO_3 + Na_3PO_4 \rightarrow$ _____

   5. $Cu(NO_3)_2 + Na_3PO_4 \rightarrow$ _____

   6. $Al(NO_3)_3 + Na_3PO_4 \rightarrow$ _____

E. Neutralization Reactions

   1. $HNO_3 + NaOH \rightarrow$ _____

   2. $H_2SO_4 + NaOH \rightarrow$ _____

   3. $H_3PO_4 + NaOH \rightarrow$ _____

A. Combination Reactions—Instructor Demonstrations

1. magnesium(s) + oxygen(g) $\xrightarrow{\Delta}$ magnesium oxide(s)

   $Mg(s)$ + $O_2(g)$ $\xrightarrow{\Delta}$

2. sulfur(s) + oxygen(g) $\xrightarrow{\Delta}$ sulfur dioxide(g)

   $S(s)$ + $O_2(g)$ $\xrightarrow{\Delta}$

B. Decomposition Reactions

1. copper(II) sulfate pentahydrate(s) $\xrightarrow{\Delta}$ copper(II) sulfate(s) + water(g)

   $CuSO_4 \cdot 5H_2O(s)$ $\xrightarrow{\Delta}$

2. sodium hydrogen carbonate(s) $\xrightarrow{\Delta}$ sodium carbonate(s) + water(g) + carbon dioxide(g)

   $NaHCO_3(s)$ $\xrightarrow{\Delta}$

C. Single-Replacement Reactions

1. copper(s) + silver nitrate(aq) $\rightarrow$ copper(II) nitrate(aq) + silver(s)

   $Cu(s)$ + $AgNO_3(aq)$ $\rightarrow$

2. magnesium(s) + hydrochloric acid(aq) $\rightarrow$ magnesium chloride(aq) + hydrogen(g)

   $Mg(s)$ + $HCl(aq)$ $\rightarrow$

3. calcium(s) + water(l) $\rightarrow$ calcium hydroxide(s) + hydrogen(g)

   $Ca(s)$ + $H_2O(l)$ $\rightarrow$

D. Double-Replacement Reactions

1. silver nitrate(aq) + potassium carbonate(aq) → silver carbonate(s) + potassium nitrate(aq)

   $AgNO_3(aq)$ + $K_2CO_3(aq)$ →

2. copper(II) nitrate(aq) + potassium carbonate(aq) → copper(II) carbonate(s) + potassium nitrate(aq)

   $Cu(NO_3)_2(aq)$ + $K_2CO_3(aq)$ →

3. aluminum nitrate(aq) + potassium carbonate(aq) → aluminum carbonate(s) + potassium nitrate(aq)

   $Al(NO_3)_3(aq)$ + $K_2CO_3(aq)$ →

4. silver nitrate(aq) + sodium phosphate(aq) → silver phosphate(s) + sodium nitrate(aq)

   $AgNO_3(aq)$ + $Na_3PO_4(aq)$ →

5. copper(II) nitrate(aq) + sodium phosphate(aq) → copper(II) phosphate(s) + sodium nitrate(aq)

   $Cu(NO_3)_2(aq)$ + $Na_3PO_4(aq)$ →

6. aluminum nitrate(aq) + sodium phosphate(aq) → aluminum phosphate(s) + sodium nitrate(aq)

   $Al(NO_3)_3(aq)$ + $Na_3PO_4(aq)$ →

E. Neutralization Reactions

1. nitric acid(aq) + sodium hydroxide(aq) → sodium nitrate(aq) + water(l)

   $HNO_3(aq)$ + $NaOH(aq)$ →

2. sulfuric acid(aq) + sodium hydroxide(aq) → sodium sulfate(aq) + water(l)

   $H_2SO_4(aq)$ + $NaOH(aq)$ →

3. phosphoric acid(aq) + sodium hydroxide(aq) → sodium phosphate(aq) + water(l)

   $H_3PO_4(aq)$ + $NaOH(aq)$ →

F. Percentages of Copper and Zinc in a Penny   **Mint Date** _____

      mass of "zinc penny"   _____ g

      mass of copper   _____ g

      mass of zinc   _____ g

Show the calculation for the percentage of copper in the penny (see Example Exercise 10.1).

      Percentage of copper   _____ %

Show the calculation for the percentage of zinc in the penny.

      Percentage of zinc   _____ %

POSTLABORATORY ASSIGNMENT

1. Provide the chemical formula for the substance described in each of the chemical reactions. Refer to pages 122–123 for the substance described in each reaction.

   (a) the white smoke produced from reaction **A.1**   _____

   (b) the pungent odor produced from reaction **A.2**   _____

   (c) the colorless gas produced from reaction **B.1**   _____

   (d) the flame-extinguishing gas from reaction **B.2**   _____

   (e) the gray solid produced from reaction **C.1**   _____

   (f) the colorless gas produced from reaction **C.2**   _____

   (g) the white ppt produced from reaction **C.3**   _____

   (h) the cream ppt produced from reaction **D.1**   _____

   (i) the blue-white ppt produced from reaction **D.2**   _____

   (j) the white ppt produced from reaction **D.3**   _____

   (k) the yellow ppt produced from reaction **D.4**   _____

   (l) the blue-white ppt produced from reaction **D.5**   _____

   (m) the white ppt produced from reaction **D.6**   _____

   (n) the acid reacting in reaction **E.1**   _____

   (o) the acid reacting in reaction **E.2**   _____

   (p) the acid reacting in reaction **E.3**   _____

   (q) the base reacting in reactions **E.1– E.3**   _____

2. Refer to the *Activity Series* in **Appendix G** and indicate reaction (*Rxn*) or no reaction (*NR*) when a small piece of zinc metal is dropped into the following aqueous solutions.

   (a) $Mn(NO_3)_2(aq)$ _____   (b) $Fe(NO_3)_3(aq)$ _____

   (c) $LiNO_3(aq)$ _____   (d) $AgNO_3(aq)$ _____

   (e) $Al(NO_3)_3(aq)$ _____   (f) $HNO_3(aq)$ _____

3. Refer to the *Solubility Rules for Ionic Compounds* in **Appendix H** and indicate whether the following ionic compounds are soluble (*sol*) or insoluble (*insol*) in water.

(a) $(NH_4)_2CO_3$ _____

(b) $Mg(C_2H_3O_2)_2$ _____

(c) $Fe(NO_3)_3$ _____

(d) Hg_2Cl_2 _____

(e) $ZnSO_4$ _____

(f) $CaCO_3$ _____

(g) $PbCrO_4$ _____

(h) K_3PO_4 _____

(i) CuS _____

(j) $Ba(OH)_2$ _____

4. Convert the following word equations into balanced chemical equations.

(a) copper metal(s) + chlorine(g) $\xrightarrow{\Delta}$ copper(II) chloride(s)

(b) manganese(II) carbonate(s) $\xrightarrow{\Delta}$ manganese(II) oxide(s) + carbon dioxide(g)

(c) potassium metal(s) + water(l) → potassium hydroxide(aq) + hydrogen(g)

(d) barium chloride(aq) + sodium sulfate(aq) → barium sulfate(s) + sodium chloride(aq)

(e) acetic acid(aq) + aluminum hydroxide(s) → aluminum acetate(aq) + water(l)

5. (optional) A 1976 penny weighing 3.080 g is dissolved in nitric acid, and copper metal is electroplated from the solution. If the mass of copper electroplated from solution is 2.926 g, what is the percentage of copper in the 1976 penny?

_____ % Cu

Determination of Avogadro's Number

- To determine Avogadro's number using a molecular monolayer technique.
- To find the number of molecules in a monolayer.
- To find the moles of stearic acid in a monolayer.
- To develop a sensitive technique in preparing a thin film of molecules.

Avogadro's number (symbol N) is defined as the number of carbon atoms in 12.01 g of carbon. It is an extremely large number. Avogadro's number has been determined by several experimental methods. Currently, the most precise known value is 6.0221367×10^{23}.

A **mole** (symbol **mol**) is the amount of substance that contains Avogadro's number of atoms, molecules, or formula units. If we use a value of 6.02×10^{23} for Avogadro's number, we can write

$$\text{Avogadro's number (N)} \quad = \quad 1 \text{ mole} \quad = \quad 6.02 \times 10^{23} \text{ particles}$$

If we refer to the periodic table, we find that the atomic mass of carbon is 12.01 amu. Therefore, 1 mole of carbon, 6.02×10^{23} C atoms, has a mass equal to 12.01 g. Similarly, 1 mole of any other element has a mass equal to its atomic mass expressed in grams; for example, 1 mole of aluminum, 6.02×10^{23} Al atoms, has a mass of 26.98 g.

The mass of one mole of any substance is termed its **molar mass** (symbol **MM**). The molar mass of carbon is 12.01 g, the molar mass of oxygen, O_2, is 32.00 g, and the molar mass of sodium chloride, NaCl, is 58.44 g. Since we know the mass of one mole of any substance we also know the number of atoms corresponding to its molar mass. Thus,

$$6.02 \times 10^{23} \text{ particles} \quad = \quad 1 \text{ mole} \quad = \quad \text{molar mass (g/mol)}$$

In this experiment, we will determine an experimental value for Avogadro's number. First, a solution of stearic acid is prepared by dissolving the solid substance in the organic solvent hexane. The solution is then added one drop at a time onto the water contained in a watchglass. After each drop is added, the stearic acid molecules spread across the surface of the water forming a single layer. This single layer of molecules is referred to as a **monolayer**. A few seconds after each drop of solution is added, the hexane solvent evaporates and the drop disappears. When enough drops of solution have been added to form a monolayer of stearic acid molecules, *one* additional drop forms a clear bead or lens (Figure 11.1).

clear lens

1 excess drop of solution

stearic acid monolayer on water

Figure 11.1 Formation of a Molecular Monolayer A single monolayer of stearic acid molecules spreads across the surface of the water in a watchglass.

The stearic acid molecule, $C_{17}H_{35}COOH$, is a long-chain molecule having a polar "head" and a nonpolar "tail." The structure of the molecule is as follows:

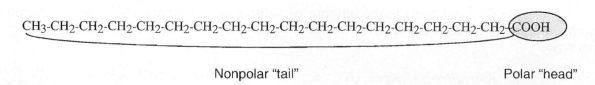

$CH_3\text{-}CH_2\text{-}CH_2\text{-}CH_2\text{-}CH_2\text{-}CH_2\text{-}CH_2\text{-}CH_2\text{-}CH_2\text{-}CH_2\text{-}CH_2\text{-}CH_2\text{-}CH_2\text{-}CH_2\text{-}CH_2\text{-}CH_2\text{-}CH_2\text{-}COOH$

Nonpolar "tail"   Polar "head"

The polar "head" is soluble in water, which is a polar liquid. The nonpolar "tail" is insoluble in water, just as oil is insoluble in water. The monolayer is composed of stearic acid molecules that have their polar "heads" dissolved in the water and their nonpolar "tails" repelled from the surface of the water (Figure 11.2).

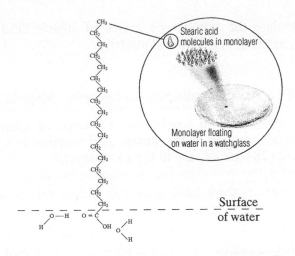

Figure 11.2 Enlarged View of Stearic Acid in Water The polar –COOH "head" dissolves in water, but the nonpolar "tail" of the molecule is insoluble.

The area occupied by all the molecules in the monolayer is called the **surface area**. Assuming that there are no spaces between stearic acid molecules in the monolayer, each molecule occupies an area of about 0.21 nm^2. The following example exercise illustrates the calculation for an experimental value of Avogadro's number.

Example Exercise 11.1 • Calculating Molecules in the Monolayer

A stearic acid solution dropped onto water gave a monolayer with a diameter of 14.5 cm. Calculate the number of stearic acid molecules in the monolayer, assuming each molecule occupies an area of 0.21 nm^2.

Solution: First, we must find the surface area of the monolayer. We can calculate the area of a circle using the formula πr^2. Since the radius (r) is half the diameter (d), we can write the formula for the circular monolayer as:

$$\frac{\pi d^2}{4} \quad = \quad \text{surface area}$$

$$\frac{(3.14)(14.5 \text{ cm})^2}{4} \quad = \quad 165 \text{ cm}^2$$

Second, we can find the number of molecules by dividing the surface area of the monolayer (165 cm^2) by the area of a single molecule (0.21 nm^2). Since the areas are expressed in different units, we will convert cm^2 to nm^2.

$$\frac{165 \text{ cm}^2}{\text{monolayer}} \times \left(\frac{1 \text{ m}}{1 \infty 10^2 \text{ cm}}\right)^2 \times \left(\frac{1 \infty 10^9 \text{ nm}}{1 \text{ m}}\right)^2 \times \frac{1 \text{ molecule}}{0.21 \text{ nm}^2} = \frac{\text{molecules}}{\text{monolayer}}$$

$$\frac{165 \text{ cm}^2}{\text{monolayer}} \times \frac{1 \text{ m}^2}{1 \infty 10^4 \text{ cm}^2} \times \frac{1 \infty 10^{18} \text{ nm}^2}{1 \text{ m}^2} \times \frac{1 \text{ molecule}}{0.21 \text{ nm}^2} = \frac{\text{molecules}}{\text{monolayer}}$$

$$= 7.9 \times 10^{16} \text{ molecules/monolayer}$$

In this example, the monolayer contains 7.9×10^{16} molecules of stearic acid.

Next, we can calculate the number of moles of stearic acid in the monolayer. Let's continue with the determination of Avogadro's number.

Example Exercise 11.2 • Calculating Moles in the Monolayer

A calibrated pipet (65 drops/mL) delivered 11 drops of stearic acid solution onto the monolayer. If the concentration of the stearic acid solution is 1.4×10^{-4} g/mL, how many moles of stearic acid (284 g/mol) are in the monolayer?

Solution: The moles of stearic acid in the monolayer are calculated as follows:

$$\frac{11 \text{ drops}}{\text{monolayer}} \times \frac{1 \text{ mL solution}}{65 \text{ drops}} \times \frac{1.4 \times 10^{-4} \text{ g}}{1 \text{ mL solution}} \times \frac{1 \text{ mol}}{284 \text{ g}} = \frac{\text{mol}}{\text{monolayer}}$$

$$= 8.3 \times 10^{-8} \text{ mol/monolayer}$$

In this example, the monolayer contains 8.3×10^{-8} moles of stearic acid.

Finally, we can calculate a value for Avogadro's number by comparing the molecules of stearic acid in the monolayer to the moles of stearic acid in the monolayer.

Example Exercise 11.3 • Determining Avogadro's Number

If the monolayer contains 7.9×10^{16} molecules of stearic acid and 8.3×10^{-8} moles, what is the experimental value for Avogadro's number?

Solution: The experimental value for Avogadro's number is simply the ratio of the number of stearic acid molecules per mole:

$$\text{Avogadro's number} = \frac{7.9 \times 10^{16} \text{ molecules}}{8.3 \times 10^{-8} \text{ mol}}$$

$$= 9.5 \times 10^{23} \text{ molecules/mol}$$

The experimental value for Avogadro's number is 9.5×10^{23}. This value is typical for the experiment, which tends to give somewhat high results.

- dropper pipet (or 15 cm of 6 mm glass tubing)
- dropper pipet bulb
- 13 × 100 mm test tube
- 15 cm watchglass
- wash bottle with distilled water
- metric ruler
- stearic acid solution
  (0.00012–0.00015 g/mL of hexane is recommended)
- 10-mL graduated cylinder

PROCEDURE

A. Calibrating a Dropper Pipet

1. Cut a 15 cm length of 6 mm glass tubing. Heat the tubing and draw it into a fine capillary tip.

   Note: A commercial dropper pipet may be substituted; however, for best results the dropper calibration should exceed 50 drops/mL.

2. Obtain about 3 mL of stearic acid solution in a test tube. Calibrate the dropper pipet by adding drops of stearic acid solution into a 10-mL graduated cylinder. Hold the dropper at a 45° angle, and deliver the drops at a rate of one per second. Record the number of drops to reach the 1 mL mark.

3. Repeat the dropwise calibration procedure twice and find the average number of drops for the three trials.

B. Calculating Molecules in the Monolayer

1. Measure the diameter of a large watchglass to 0.1 cm. Record the diameter, and calculate the surface area of the monolayer; that is, the surface area of the watchglass.

   Note: The diameter of a stearic acid monolayer corresponds to the surface area of the watchglass filled with water.

2. Calculate the number of molecules that can occupy a monolayer assuming the area of a stearic acid molecule is 0.21 nm^2.

C. Determining Avogadro's Number

1. Clean the watchglass carefully with soap and water. Rinse the watchglass thoroughly with distilled water, and do not touch the inside concave surface.

2. Place the convex side of the watchglass on a paper towel on the lab bench. Fill the watchglass completely with distilled water from a wash bottle.

3. Record the concentration of the stearic acid solution. Hold the dropper pipet at a 45° angle, and slowly deliver drops of solution onto the center of the water surface. Continue adding drops of solution until a clear lens persists for at least 30 seconds.

   Note: A small, persistent, clear lens indicates a monolayer of molecules has formed across the entire surface of the water.

4. In the Data Table, record the number of drops required to form the monolayer.

5. Find the moles of stearic acid in the monolayer, and determine the experimental value for Avogadro's number.

6. Thoroughly clean and rinse the watchglass and perform a second trial.

7. Thoroughly clean and rinse the watchglass and perform a third trial.

EXPERIMENT 11 NAME ____________________

DATE ____________________ SECTION ____________________

*PRELABORATORY ASSIGNMENT**

1. In your own words, define the following terms:

   Avogadro's number (N)

   molar mass (MM)

   mole (mol)

   monolayer

   surface area

2. Circle the portion of the stearic acid molecule shown below that is polar and dissolves in water.

   CH_3-CH_2-CH_2-CH_2-CH_2-CH_2-CH_2-CH_2-CH_2-CH_2-CH_2-CH_2-CH_2-CH_2-CH_2-CH_2-CH_2–COOH

3. What is observed after the hexane solvent evaporates?

4. When do you stop adding drops of stearic acid solution to the monolayer?

5. What are the major sources of error in this experiment?

** Answers in Appendix J*

6. A calibrated pipet (95 drops/mL) delivers 12 drops of stearic acid solution (1.5×10^{-4} g/mL) and produces a monolayer with a diameter of 12.5 cm. Calculate

   (a) the surface area of the monolayer (refer to Example Exercise 11.1).

   (b) the number of stearic acid molecules in the monolayer assuming each molecule occupies an area of 0.21 nm^2 (refer to Example Exercise 11.1).

   (c) the moles of stearic acid (284 g/mol) in the monolayer (refer to Example Exercise 11.2).

   (d) the experimental value for Avogadro's number (refer to Example Exercise 11.3).

7. What safety precautions must be observed in this experiment?

DATE _____

NAME _____

SECTION _____

DATA TABLE

A. Calibrating a Dropper Pipet

number of drops per milliliter   _____   _____   _____

      Average value   _____ drops/mL

B. Calculating Molecules in the Monolayer

diameter of monolayer *(diameter of watchglass)*   _____ cm

area of one stearic acid molecule   __0.21__ nm^2

Show the calculation for the surface area of the monolayer (see Example Exercise 11.1).

surface area of monolayer   _____ cm^2

Show the calculation for the number of stearic acid molecules in the monolayer.

molecules in monolayer   _____ molecules

C. Determining Avogadro's Number

number of drops in monolayer _____   _____   _____

concentration of stearic acid solution   _____ g/mL
(see Instructor)

Show the calculation for the moles of stearic acid in the monolayer for trial 1 (see Example Exercise 11.2).

moles in monolayer   _____ mol _____ mol _____ mol

Show the calculation for the number of molecules of stearic acid per mole for trial 1 (see Example Exercise 11.3).

Avogadro's number *(N)*   _____   _____   _____

Average value for *N*   _____ molecules/mol

EXPERIMENT 11 NAME ______________________

DATE ______________ SECTION ______________________

POSTLABORATORY ASSIGNMENT

1. A solution of oleic acid, $C_{17}H_{33}COOH$, is added one drop at a time onto water until a monolayer is formed. Assume that there are no spaces between molecules in the monolayer and that each oleic acid molecule occupies an area of 0.25 nm^2. If the concentration of the oleic acid solution is 0.00012 g/mL, what is the experimental value of Avogadro's number?

dropper pipet calibration	=	68 drops/mL
number of drops of oleic acid in the monolayer	=	16 drops
diameter of monolayer	=	15.0 cm

N = ______________

2. Calculate the mass of carbon in a 1-carat diamond that contains 1.00×10^{22} atoms of carbon.

____

3. Calculate the mass of chloroform that contains 1,000,000,000,000 molecules of $CHCl_3$.

_____

4. Calculate the number of formula units in 1.37 g of NaCl; that is, table salt.

_____

5. Calculate the number of N_2 molecules in 0.111 g of nitrogen gas.

_____

6. What is the mass of one molecule of stearic acid, $C_{17}H_{35}COOH$, expressed in grams?

_____

7. (optional) Earth's oceans have a surface area of 360,000,000 km^2, and a ping-pong ball floating on water occupies a surface area of about 15 cm^2. Assuming there is no space between ping-pong balls, how many planets similar to Earth would be necessary to float Avogadro's number of ping-pong balls?

_____

Empirical Formulas of Compounds

- To determine the empirical formula for magnesium oxide.
- To determine the empirical formula for copper sulfide.
- To gain practical experience in developing techniques using a crucible.

During the late 1700s, chemists experimented with elements to see how they reacted to form compounds. In particular, they were interested in the reactions of metals as they combined with oxygen gas in the air. By measuring the mass of a metal before reaction and the mass of the metal oxide after reaction, chemists were able to determine the formulas of metal oxide compounds.

The simplest whole number ratio of atoms in a compound is referred to as the **empirical formula**. Originally, an element was placed in a particular group in the periodic table based on the empirical formula of its oxide. For example, magnesium, calcium, strontium, and barium were placed in Group IIA/2 because they react with oxygen to give similar empirical formulas; that is, MgO, CaO, SrO, and BaO. Moreover, the empirical formulas of their chlorides are also similar; that is, $MgCl_2$, $CaCl_2$, $SrCl_2$, and $BaCl_2$.

Since transition metals can combine with nonmetals in different ratios, we cannot always predict the empirical formulas of their compounds. For example, iron can combine with oxygen to form either iron(II) oxide, FeO, or iron(III) oxide, Fe_2O_3. The following example exercises illustrate the calculation of empirical formulas.

139

Example Exercise 12.1 • Determining an Empirical Formula

A 0.279-g sample of iron is heated and allowed to react with oxygen from the air. If the product has a mass of 0.400 g, what is the empirical formula of the iron oxide?

Solution: The empirical formula is experimentally determined from the moles of each reactant. The moles of iron are calculated as follows.

$$0.279 \; \cancel{\text{g Fe}} \; \times \; \frac{1 \text{ mol Fe}}{55.85 \; \cancel{\text{g Fe}}} \; = \; 0.00500 \text{ mol Fe}$$

The mass of oxygen that reacted is 0.400 g product – 0.279 g iron = 0.121 g. We can calculate the moles of oxygen as follows.

$$0.121 \; \cancel{\text{g O}} \; \times \; \frac{1 \text{ mol O}}{16.00 \; \cancel{\text{g O}}} \; = \; 0.00756 \text{ mol O}$$

The mole ratio of the elements in iron oxide is $Fe_{0.00500}O_{0.00756}$, and we can divide by 0.00500 to find the simplest whole number ratio.

$$Fe \frac{0.00500}{0.00500} \; O \frac{0.00756}{0.00500} \; = \; Fe_{1.00}O_{1.51}$$

If we double the mole ratio, we obtain $Fe_2O_{3.02}$. We can explain the slight deviation from a whole number by experimental error. The empirical formula is Fe_2O_3, and we name the compound iron(III) oxide, or ferric oxide.

Example Exercise 12.2 • Determining an Empirical Formula

A 0.331-g sample of iron is placed in a crucible and covered with powdered sulfur. The crucible is heated until all the excess sulfur is driven off. If the product weighs 0.522 g, what is the empirical formula of the iron sulfide?

Solution: First, we can calculate the moles of iron in the product.

$$0.331 \; \cancel{\text{g Fe}} \; \times \; \frac{1 \text{ mol Fe}}{55.85 \; \cancel{\text{g Fe}}} \; = \; 0.00592 \text{ mol Fe}$$

The mass of sulfur that reacted is 0.522 g product – 0.331 g iron = 0.191 g. Second, we can calculate the moles of sulfur as follows.

$$0.191 \; \cancel{\text{g S}} \; \times \; \frac{1 \text{ mol S}}{32.07 \; \cancel{\text{g S}}} = 0.00596 \text{ mol S}$$

The mole ratio of the elements in iron sulfide is $Fe_{0.00592}S_{0.00596}$, and we divide by 0.00592 to find the simplest whole number ratio.

$$Fe \frac{0.00592}{0.00592} \; S \frac{0.00596}{0.00592} \; = \; Fe_{1.00}S_{1.01}$$

We explain the slight deviation from whole numbers by experimental error. The empirical formula for the product is FeS, and we name the compound iron(II) sulfide, or ferrous sulfide.

In this experiment, you will ignite magnesium ribbon in a crucible and convert the metal to an oxide product. The second part of the experiment involves the conversion of copper to copper sulfide. Since copper can form either copper(I) sulfide or copper(II) sulfide, the empirical formula is unknown and cannot be predicted. Figure 12.1 illustrates the experimental equipment.

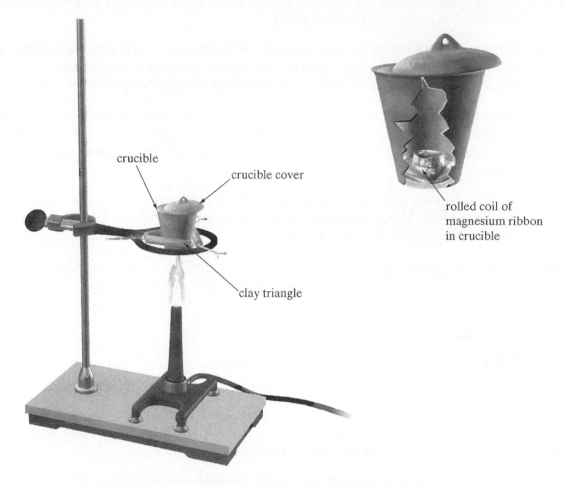

Figure 12.1 Empirical Formula Apparatus A crucible and cover are placed in a clay triangle on a ring stand, and heated until red hot.

EQUIPMENT and CHEMICALS

- clay triangle
- crucible tongs
- crucible and cover

- magnesium, Mg ribbon
- copper, #18 gauge Cu wire
- sulfur, S powder

PROCEDURE

A. Empirical Formula of Magnesium Oxide

1. Support a crucible and cover with a clay triangle, and place on a ring stand. Fire the crucible and cover to red heat using the tip of the flame from a laboratory burner.

2. Turn off the burner, and allow the crucible and cover to cool for 10 minutes. Weigh the crucible, and cover on a balance.

3. Cut a 25-cm strip of magnesium ribbon, and roll the metal into a flat coil. Place the coil of magnesium in the crucible so that it lies flat against the bottom. Reweigh the crucible, cover, and magnesium metal.

4. Return the crucible to the clay triangle. With the cover off, fire the crucible to red heat. When the magnesium sparks and begins to smoke, immediately remove the burner and place the cover on the crucible using crucible tongs.

5. After the smoke has ceased, continue to heat the crucible and cover until the metal is completely converted to a gray-white residue. The progress of the reaction can be checked periodically by removing the burner and raising the cover with the crucible tongs.

6. When the metal no longer sparks, turn off the burner and allow the crucible to cool for 10 minutes. Using a dropper pipet, add drops of distilled water until no fizzing is observed from the gray-white residue.

7. Cover the crucible, and heat for 5 minutes. Turn off the burner, and allow the crucible to cool for 10 minutes. Weigh the crucible and cover containing the magnesium oxide.

8. Clean the crucible, and repeat the procedure.

9. Calculate the empirical formula for each trial.

B. Empirical Formula of Copper Sulfide

> **Caution:** This procedure requires a vented fume hood, as burning sulfur produces pungent sulfur dioxide gas.

1. Support a crucible and cover with a clay triangle and place on a ring stand. Fire the crucible and cover to red heat.

2. Remove the heat, and allow the crucible and cover to cool for 10 minutes. Weigh the crucible and cover.

3. Cut a 25-cm length of copper wire, and roll the wire into a coil. Place the wire in the bottom of the crucible, and reweigh the crucible, cover, and copper wire.

4. Cover the copper wire completely with powdered sulfur. Place the cover on the crucible, and gradually heat to red heat under a fume hood. Continue to heat for several minutes after the last trace of burning sulfur disappears. Hold the burner in your hand and continue to heat the entire outside surface of the crucible and cover.

5. Allow the crucible and contents to cool for 10 minutes. Weigh the crucible and cover containing the copper sulfide.

6. Clean the crucible, and repeat the procedure.

7. Calculate the empirical formula for each trial.

PRELABORATORY ASSIGNMENT*

1. In your own words, define the following terms:

    empirical formula

    firing to red heat

    heating to constant weight

    molecular formula

    weighing by difference

2. Why are the empty crucible and cover fired to red heat?

3. How critical are the suggested times for heating and cooling?

4. Why is distilled water added to the crucible after igniting the magnesium metal?

5. How can you tell when the magnesium metal has reacted completely?

6. How can you tell when the copper wire has reacted completely and the excess sulfur has burned off?

---

* Answers in Appendix J

7. A 0.250-g sample of calcium metal is heated in a 38.500-g crucible to form calcium oxide. The resulting crucible and product weigh 38.850 g. Refer to Example Exercise 12.1 and determine the empirical formula of calcium oxide.

8. What are the major sources of error in this experiment?

9. What safety precautions must be observed in this experiment?

NAME _____

SECTION _____

DATA TABLE

A. Empirical Formula of Magnesium Oxide

mass of crucible and cover + magnesium metal
(*before heating*)   _____ g   _____ g

mass of crucible and cover   _____ g   _____ g

mass of magnesium metal   _____ g   _____ g

mass of crucible and cover + magnesium oxide
(*after heating*)   _____ g   _____ g

mass of combined oxygen   _____ g   _____ g
(*after heating – before heating*)

Show the calculation of the empirical formula for trial 1 (see Example Exercise 12.1).

Empirical formula of magnesium oxide   _____   _____

B. Empirical Formula of Copper Sulfide

mass of crucible and cover + copper wire
(before heating) _____ g   _____ g

mass of crucible and cover _____ g   _____ g

mass of copper wire _____ g   _____ g

mass of crucible and cover + copper sulfide
(after heating) _____ g   _____ g

mass of crucible and cover + copper sulfide
(optional second heating) _____ g   _____ g

mass of combined sulfur _____ g   _____ g
(after heating – before heating)

Show the calculation of the empirical formula for trial 1 (see Example Exercise 12.2).

Empirical formula of copper sulfide _____   _____

NAME _____ *ANSWER KEY* _____

DATE _____

SECTION _____

1. Predict the empirical formula for each of the following compounds given the formula of magnesium oxide, MgO.

    (a) calcium oxide   _____   (b) strontium oxide   _____

    (c) barium oxide   _____   (d) radium oxide   _____

2. Predict the empirical formula for each of the following compounds given the formula of sodium chloride, NaCl.

    (a) sodium fluoride   _____   (b) potassium chloride   _____

    (c) lithium bromide   _____   (d) rubidium iodide   _____

3. A 1.250-g sample of copper wire was heated in air and reacted with oxygen to give 1.565 g of copper oxide product. Calculate the empirical formula of the copper oxide.

    _____

4. A 1.167-g sample of iron powder reacted with 0.680 g of sulfur powder to give iron sulfide product. Calculate the empirical formula of the iron sulfide.

    _____

5. A 0.565-g sample of cobalt metal reacted with excess sulfur powder to give 1.027 g of cobalt sulfide. Calculate the empirical formula of the product.

_____

6. A 0.750-g sample of tin foil was heated in air and reacted with oxygen gas to give 0.953 g of tin oxide. Calculate the empirical formula of the product.

_____

7. (optional) A 1.000-g sample of red phosphorus powder was burned in air and reacted with oxygen gas to give 2.291 g of phosphorus oxide. Calculate the empirical formula and molecular formula of phosphorus oxide given the molar mass is approximately 282 g/mol?

Empirical formula _____

Molecular formula _____

Analysis of Alum

- To determine the percentage of water in alum hydrate.
- To determine the percentage of water in an unknown hydrate.
- To calculate the water of crystallization for an unknown hydrate.
- To develop the laboratory skills for analyzing a hydrate.

A **hydrate** is a compound having a fixed number of water molecules. The number of water molecules is referred to as the **water of crystallization**, or water of hydration. For example, barium chloride dihydrate, $BaCl_2 \cdot 2H_2O$, has two waters of crystallization; and alum hydrate, $KAl(SO_4)_2 \cdot 12H_2O$, has twelve waters of hydration. When heated, a hydrate loses water and produces an **anhydrous** compound. Alum, a compound used in styptic pencils to stop bleeding of minor cuts, is decomposed as follows:

$$KAl(SO_4)_2 \cdot 12H_2O(s) \xrightarrow{\Delta} KAl(SO_4)_2(s) + 12\ H_2O(g)$$

alum	*anhydrous*	*water of*
hydrate	*compound*	*crystallization*

The theoretical percentage of water in a hydrate is found by comparing the mass of the water of crystallization to the mass of the hydrate. This is accomplished by dividing the total mass of water by the molar mass of the hydrate. Example Exercise 13.1 illustrates this calculation.

Example Exercise 13.1 • Theoretical % H₂O in Alum

Calculate the theoretical percentage of water in alum hydrate, $KAl(SO_4)_2 \cdot 12H_2O$.

Solution: The molar mass of $KAl(SO_4)_2 \cdot 12H_2O$ is found as follows:

1	×	39.10 g	=	39.10 g
1	×	26.98 g	=	26.98 g
2	×	96.07 g	=	192.14 g
12	×	18.02 g	=	216.24 g
				474.46 g

The theoretical percentage of water is found by dividing the total mass of water (12×18.02 g) by the molar mass of the hydrate (474.46 g).

$$\frac{12 \times 18.02 \text{ g}}{474.46 \text{ g}} \times 100\% = 45.58\% \text{ H}_2\text{O}$$

Experimentally, the amount of water is found from **weighing by difference**; that is, the difference in mass before and after heating the sample. For example, an alum hydrate sample is decomposed to give the following data:

mass of beaker and watchglass + alum hydrate	102.636 g
mass of beaker and watchglass	101.486 g
mass of beaker and watchglass + anhydrous compound	102.113 g

The mass of the hydrate is found by subtraction: 102.636 g – 101.486 g = 1.150 g. The mass of the water driven off by decomposition is equal to the difference before and after heating the sample: 102.636 g – 102.113 g = 0.523 g. The calculation for the experimental percentage of water in the alum sample is illustrated in Example Exercise 13.2.

Example Exercise 13.2 • Experimental % H₂O in Alum

A sample of alum has a mass of 1.150 g and loses 0.523 g of water when decomposed by heating. Calculate the experimental percentage of water in alum hydrate.

Solution: The experimental percentage of water is simply

$$\frac{\text{mass of water}}{\text{mass of hydrate}} \times 100\% = \% \text{ water}$$

$$\frac{0.523 \text{ g}}{1.150 \text{ g}} \times 100\% = 45.5\% \text{ water}$$

The proficiency of your laboratory technique can be evaluated by comparing the experimental percentage of water to the theoretical value. In the previous examples, the two values are in good agreement. That is, the theoretical percentage of water in $KAl(SO_4)_2 \cdot 12H_2O$ is 45.58%, and the experimental value was found to be 45.5%.

In this experiment, a sample of alum hydrate will be heated to determine the percentage of water (Figure 13.1). After gaining experience analyzing the known alum hydrate, you will analyze an unknown hydrate.

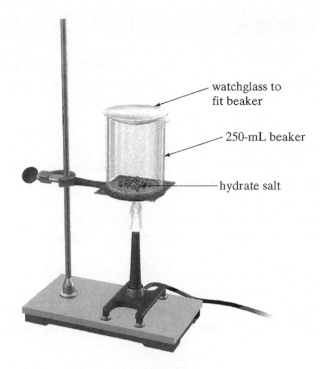

watchglass to
fit beaker

250-mL beaker

hydrate salt

Figure 13.1 Apparatus for Decomposing a Hydrate A hydrate compound is
heated gently to release steam and to avoid spattering.

After analyzing an alum sample, you will analyze an unknown hydrate for the percentage
of water using the same procedure. You will also determine the water of crystallization for the
unknown hydrate; that is, you will determine the value of X in an unknown hydrate, $AC \cdot XH_2O$.

Example Exercise 13.3 • Water of Crystallization for Unknown Hydrate

Calculate the water of crystallization for an unknown hydrate, $AC \cdot XH_2O$, that is found to
contain 30.6% water. The molar mass of the anhydrous compound (AC) is 245 g/mol.

Solution: If the amount of water in the unknown hydrate is 30.6%, the anhydrous
compound must be 69.4% (100% − 30.6% = 69.4%). In a 100.0-g sample, the
mass of water is 30.6 g and the mass of anhydrous compound is 69.4 g. We
can calculate the moles of water and anhydrous compound as follows.

$$30.6 \; \text{g} \, H_2O \; \times \; \frac{1 \text{ mol } H_2O}{18.02 \; \text{g} \, H_2O} \; = \; 1.70 \text{ mol } H_2O$$

$$69.4 \; \text{g} \, AC \; \times \; \frac{1 \text{ mol AC}}{245 \; \text{g} \, AC} \; = \; 0.283 \text{ mol AC}$$

To find the water of crystallization we simply divide the moles of water by
the moles of anhydrous compound. Thus,

$$\frac{1.70 \text{ mol } H_2O}{0.283 \text{ mol AC}} \; = \; 6.01 \approx 6$$

Since the water of crystallization must be a whole number, we round off 6.01
to the nearest whole number (6). The formula of the hydrate is $AC \cdot 6H_2O$.

- wire gauze
- 250-mL beaker
- watchglass

- alum, $KAl(SO_4)_2 \cdot 12H_2O$
- unknown hydrate samples

PROCEDURE

A. Percentage of Water in Alum Hydrate

1. Weigh a clean, dry, 250-mL beaker covered with a watchglass. Add about 0.8–1.2 g of alum hydrate into the beaker, and reweigh accurately.

2. Support the beaker and watchglass on a ring stand using a wire gauze (Figure 13.1). Heat the hydrate gently, and you should observe moisture on the sides of the beaker and bottom of the watchglass. Continue heating until all the moisture is evaporated. When the hydrate is completely decomposed, it will change from crystalline to powder.

   Note: If the watchglass is not completely dry, hold it carefully with crucible tongs over a low burner flame until no moisture remains.

3. Turn off the burner, and allow the beaker to cool for 10 minutes. Carefully transfer the beaker with the watchglass to the balance. Weigh the mass of the beaker, watchglass, and anhydrous compound.

4. Discard the decomposed alum and clean the beaker. Perform a second trial with alum hydrate. Calculate the percentage of water in the hydrate for each trial and the average value.

B. Percentage of Water in an Unknown Hydrate

1. Obtain an unknown hydrate from the Instructor, and record the unknown number.

2. Repeat steps 1–4 as in Procedure A, and report the average percentage of water in the unknown hydrate.

C. Water of Crystallization in an Unknown Hydrate

1. Given the molar mass of the anhydrous compound from the Instructor, calculate the water of crystallization. The Instructor may wish to verify the experimental percentage of water before giving the molar mass of the unknown hydrate.

   Note: It is not unusual for the water of crystallization value to differ by a few tenths from a whole number; for example, 2.1 or 6.7. However, the water of crystallization must be rounded to a whole number. In this example, 2.1 is rounded to 2 and 6.7 is rounded to 7.

NAME _____

SECTION _____

1. In your own words, define the following terms:

   anhydrous

   hydrate

   molar mass (MM)

   water of crystallization

   weighing by difference

2. If you weigh a bag of microwave popcorn, before and after heating, does the bag of unpopped kernels weigh more or less than the bag of popped corn?

3. An alum hydrate sample was analyzed by decomposition and gave the following data:

mass of beaker and watchglass + alum hydrate	96.818 g
mass of beaker and watchglass	95.886 g
mass of beaker and watchglass + anhydrous compound	96.398 g

   Calculate the experimental percentage of water as shown in Example Exercise 13.2. Do the experimental results agree with the theoretical value shown in Example Exercise 13.1?

4. An unknown hydrate (AC • XH$_2$O) was analyzed and found to contain 13.7% water. Assume the molar mass of the anhydrous compound (AC) is 227 g/mol, refer to Example Exercise 13.3 and determine the water of crystallization and formula of the hydrate.

Water of crystallization _____   **Formula of hydrate**   AC • ____H$_2$O

5. How can you tell when to stop heating a hydrate because it is completely decomposed?

6. How will the weighing be affected by placing a warm beaker on the balance?

7. What are the major sources of error in this experiment?

8. What safety precautions must be observed in this experiment?

EXPERIMENT 13 NAME ______________________

DATE ______________ SECTION ______________________

DATA TABLE

A. Percentage of Water in Alum Hydrate

mass of beaker and watchglass + alum hydrate *(before heating)*	________ g	________ g
mass of beaker and watchglass	________ g	________ g
mass of alum hydrate	________ g	________ g
mass of beaker and watchglass + anhydrous compound *(after heating)*	________ g	________ g
mass of water *(before heating – after heating)*	________ g	________ g

Show the calculation for the percentage of water for trial 1 (see Example Exercise 13.2).

Percentage of water in $KAl(SO_4)_2 \cdot 12H_2O$	________ %	________ %

Average percentage of water ________ %

B. Percentage of Water in an Unknown Hydrate   **UNKNOWN #** _____

   mass of beaker and watchglass + unknown hydrate   _____ g   _____ g
   (before heating)

   mass of beaker and watchglass   _____ g   _____ g

   mass of unknown hydrate   _____ g   _____ g

   mass of beaker and watchglass + anhydrous compound _____ g   _____ g
   (after heating)

   mass of water   _____ g   _____ g
   (before heating – after heating)

Show the calculation for the percentage of water for trial 1 (see Example Exercise 13.2).

   Percentage of water in the unknown hydrate   _____ %   _____ %

      Average percentage of water   _____ %

C. Water of Crystallization in an Unknown Hydrate

   molar mass of anhydrous compound (AC)   _____ g/mol
   (see Instructor)

   percentage of water (see Procedure B)   _____ %

   percentage of anhydrous compound (AC)   _____ %

Show the calculation for the water of crystallization (see Example Exercise 13.3).

   Water of crystallization _____   **Formula of hydrate**   AC • ____H$_2$O

EXPERIMENT 13 NAME ______________________

DATE ______________________ SECTION ______________________

POSTLABORATORY ASSIGNMENT

1. Calculate the theoretical percentage of water for the following hydrates.

   (a) nickel(II) chloride dihydrate, $NiCl_2 \cdot 2H_2O$

   ____

   (b) nickel(II) chloride hexahydrate, $NiCl_2 \cdot 6H_2O$

   ____

2. If an unknown hydrate, $AC \cdot XH_2O$, has a mass of 1.555 g before heating, and a mass of 0.994 g after heating, what is the experimental percentage of water in the hydrate?

   ____

3. If the anhydrous compound (AC) in the preceding problem has a molar mass of 160 g/mol, what are the water of crystallization (*X*) and formula for the hydrate?

Water of crystallization ______ **Formula of hydrate** $AC \cdot$ ____ H_2O

4. A hydrate of zinc sulfate, $ZnSO_4 \cdot XH_2O$, decomposes to produce 43.9% water. Calculate the water of crystallization (X), and write the formula for the hydrate.

Water of crystallization _____   **Formula of hydrate** $ZnSO_4 \cdot$___H_2O

5. A hydrate of calcium sulfate, $CaSO_4 \cdot XH_2O$, contains 20.9% water. Calculate the water of crystallization (X), and write the formula of the hydrate.

Water of crystallization _____   **Formula of hydrate** $CaSO_4 \cdot$___H_2O

6. (optional) Plaster of Paris, $CaSO_4 \cdot XH_2O$, is a fractional hydrate; that is, one water molecule attaches to more than one formula unit. If a 2.002-g sample has a mass of 1.878 g after heating, find the water of crystallization (X), and write the formula for Plaster of Paris.

Water of crystallization _____   **Formula of hydrate** $CaSO_4 \cdot$___H_2O

Decomposing Baking Soda

OBJECTIVES

- To determine the percent yield of sodium carbonate from a decomposition reaction.
- To determine the percentage of sodium hydrogen carbonate in an unknown mixture.
- To gain proficiency in decomposing a compound and collecting a gas over water.

DISCUSSION

When baking soda is heated, sodium hydrogen carbonate, $NaHCO_3$, decomposes into solid sodium carbonate, while releasing steam and carbon dioxide gas. The equation for the reaction is

$$2\,NaHCO_3(s) \xrightarrow{\Delta} Na_2CO_3(s) + H_2O(g) + CO_2(g)$$

Notice that the reaction releases H_2O and CO_2 as gases but Na_2CO_3 remains a solid. If we weigh the mass of solid Na_2CO_3 produced in an experiment, the mass is referred to as the **actual yield**. Conversely, if we calculate the mass of Na_2CO_3 according to the balanced chemical equation, the mass is referred to as the **theoretical yield**.

The **percent yield** from a chemical reaction is an expression for the amount of actual yield compared to the theoretical yield. While some experimental errors lead to high results, other errors may give low results. Thus, the percent yield can be greater than—or less than—100%.

Example Exercise 14.1 • **% Yield of Na_2CO_3 from Baking Soda**

A 1.654 g sample of baking soda, $NaHCO_3$, decomposes to produce 1.028 g of solid sodium carbonate. Calculate the theoretical yield and percent yield of Na_2CO_3.

Solution: According to the balanced equation, 2 mol $NaHCO_3$ (84.01 g/mol) produce 1 mol Na_2CO_3 (105.99 g/mol). We can find the theoretical yield as follows:

$$1.654 \text{ g NaHCO}_3 \times \frac{1 \text{ mol NaHCO}_3}{84.01 \text{ g NaHCO}_3} \times \frac{1 \text{ mol Na}_2\text{CO}_3}{2 \text{ mol NaHCO}_3} \times \frac{105.99 \text{ g Na}_2\text{CO}_3}{1 \text{ mol Na}_2\text{CO}_3}$$

$$= 1.043 \text{ g Na}_2\text{CO}_3$$

Since the actual yield of Na_2CO_3 is 1.028 g, the percent yield is

$$\frac{\text{actual yield}}{\text{theoretical yield}} \times 100\% = \% \text{ yield}$$

$$\frac{1.028 \text{ g}}{1.043 \text{ g}} \times 100\% = 98.56\%$$

Percentage of Sodium Hydrogen Carbonate in an Unknown Mixture

An unknown mixture containing baking soda is decomposed using heat. The following example exercise illustrates the calculation for the percentage of baking soda in the mixture.

Example Exercise 14.2 • **% $NaHCO_3$ in an Unknown Mixture**

A 1.675 g unknown mixture containing baking soda is decomposed with heat. If the mass loss is 0.318 g, what is the percentage of baking soda, $NaHCO_3$, in the unknown mixture?

Solution: In this example, the mass loss corresponds to both the mass of water vapor and carbon dioxide gas. To simplify the calculation, we will combine $H_2O + CO_2$ into H_2CO_3 (62.03 g/mol) and rewrite the chemical equation.

$$2 \text{ NaHCO}_3(s) \xrightarrow{\Delta} \text{Na}_2\text{CO}_3(s) + \text{H}_2\text{CO}_3(g)$$

We can relate the H_2CO_3 mass loss to the mass of $NaHCO_3$ as follows:

$$0.318 \text{ g H}_2\text{CO}_3 \times \frac{1 \text{ mol H}_2\text{CO}_3}{62.03 \text{ g H}_2\text{CO}_3} \times \frac{2 \text{ mol NaHCO}_3}{1 \text{ mol H}_2\text{CO}_3} \times \frac{84.01 \text{ g NaHCO}_3}{1 \text{ mol NaHCO}_3}$$

$$= 0.861 \text{ g NaHCO}_3$$

If the sample mixture has a mass of 1.675 g, the percentage of $NaHCO_3$ is

$$\frac{\text{mass NaHCO}_3}{\text{mass sample}} \times 100\% = \% \text{ NaHCO}_3$$

$$\frac{0.861 \text{ g}}{1.675 \text{ g}} \times 100\% = 51.4\%$$

Figure 14.1 shows the experimental apparatus for decomposing baking soda, as well as the unknown baking soda mixture. As the baking soda decomposes, carbon dioxide gas is produced. The carbon dioxide gas displaces water from the Florence flask into a beaker. When the decomposition is complete, no more carbon dioxide gas is released and the water level in the beaker remains constant.

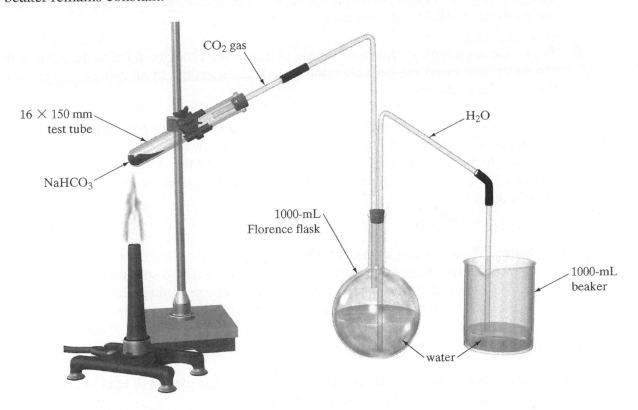

CO$_2$ gas

16 × 150 mm
test tube

NaHCO$_3$

H$_2$O

1000-mL
Florence flask

1000-mL
beaker

water

Figure 14.1 Apparatus for Decomposition When the water level in the beaker remains constant, the decomposition of NaHCO$_3$ is complete.

EQUIPMENT and CHEMICALS

- gas collection apparatus (see Figure 14.1)
- 16 × 150 mm test tube
- 1000-mL Florence flask
- 1000-mL beaker

- sodium hydrogen carbonate, baking soda, solid NaHCO$_3$
- unknown baking soda mixture, 50–90% NaHCO$_3$

A. Percent Yield of Na_2CO_3 from Baking Soda

1. Weigh a 16×150 mm *dry* test tube on the balance, and record the mass. Add 1–2 g of baking soda, $NaHCO_3$, and reweigh accurately.

2. Set up the apparatus as shown in Figure 14.1. Fill the Florence flask to the neck with tap water, and insert the gas collection apparatus. Insert the small rubber stopper into the test tube as shown.

3. Begin heating the test tube gently. Observe the water being displaced into the beaker as carbon dioxide gas is produced. As the water level in the beaker increases, continue to heat the test tube with a gentle flame. After the water level remains constant for a couple of minutes, discontinue heating and allow the test tube to cool for 10 minutes.

   Note: The decomposition of baking soda produces steam that may collect in the test tube. Any moisture in the test tube leads to serious weighing errors. If there appears to be moisture in the test tube, remove the utility clamp from the ring stand and carefully heat the open test tube over a low flame until no trace of moisture remains. Allow the test tube to cool for 10 minutes before weighing.

   Note: Avoid heating the test tube to red heat, as the glass will tend to crack and break when moisture is present.

4. Weigh the test tube containing the sodium carbonate residue. The mass of Na_2CO_3 is found by subtracting the mass of the test tube from the test tube and residue.

5. Calculate the theoretical yield of sodium carbonate, Na_2CO_3, from the mass of pure baking soda that was heated. Find the percent yield of sodium carbonate.

B. Percentage of $NaHCO_3$ in an Unknown Mixture

1. Obtain an unknown sample containing baking soda. Record the unknown number in the Data Table.

2. Repeat steps 1–5 as in Procedure A, substituting the unknown mixture for the pure baking soda.

3. Calculate the mass of baking soda, $NaHCO_3$, in the unknown sample from the mass loss. Find the percentage of baking soda in the unknown mixture.

NAME _____

DATE _____   SECTION _____

*PRELABORATORY ASSIGNMENT**

1. In your own words, define the following terms:

    actual yield

    percent yield

    stoichiometry

    theoretical yield

    weighing by difference

2. How can you tell when the baking soda sample is completely decomposed?

3. Is it possible to have a percent yield of sodium carbonate that is greater than 100%?

4. A sample of baking soda, $NaHCO_3$, is decomposed while releasing steam and carbon dioxide gas. Refer to Example Exercise 14.1 and determine the theoretical yield of sodium carbonate, Na_2CO_3, from the following data:

mass of test tube + $NaHCO_3$	17.838 g
mass of test tube	16.338 g
mass of test tube + Na_2CO_3	17.293 g

    What is the percent yield of sodium carbonate, Na_2CO_3?

_____

* Answers in Appendix J

5. An unknown mixture containing baking soda decomposes with heat and releases steam and carbon dioxide gas. Refer to Example Exercise 14.2 and find the mass of baking soda, $NaHCO_3$, in the unknown mixture from the following data:

mass of test tube + unknown mixture	18.001 g
mass of test tube	16.001 g
mass of test tube + residue	17.595 g

What is the percentage of baking soda, $NaHCO_3$, in the unknown mixture?

6. What are the major sources of error in this experiment?

7. What safety precautions must be observed in this experiment?

DATA TABLE

A. Percent Yield of Na_2CO_3 from Baking Soda

mass of test tube + $NaHCO_3$ _____ g   _____ g
(before heating)

mass of test tube _____ g   _____ g

mass of $NaHCO_3$ _____ g   _____ g

mass of test tube + Na_2CO_3 _____ g   _____ g
(after heating)

mass of Na_2CO_3 *(actual yield)* _____ g   _____ g

Show the calculation for theoretical yield of Na_2CO_3 for trial 1 (see Example Exercise 14.1).

mass of Na_2CO_3 *(theoretical yield)* _____ g   _____ g

Show the calculation for percent yield of Na_2CO_3 for trial 1 (see Example Exercise 14.1).

Percent Yield of Na_2CO_3 _____ %   _____ %

Average Percent Yield _____ %

B. Percentage of NaHCO₃ in an Unknown Mixture$\qquad$**UNKNOWN #** _____

mass of test tube + unknown mixture
(before heating)$\qquad$_____ g$\qquad$_____ g

mass of test tube$\qquad$_____ g$\qquad$_____ g

mass of unknown mixture$\qquad$_____ g$\qquad$_____ g

mass of test tube + residue
(after heating)$\qquad$_____ g$\qquad$_____ g

mass of H_2CO_3 ($H_2O + CO_2$)
(before heating – after heating)$\qquad$_____ g$\qquad$_____ g

Show the calculation for the mass of $NaHCO_3$ in the unknown mixture for trial 1 (see Example Exercise 14.2).

mass of $NaHCO_3$$\qquad$_____ g$\qquad$_____ g

Show the calculation for the percentage of $NaHCO_3$ in the unknown mixture for trial 1 (see Example Exercise 14.2).

Percentage of $NaHCO_3$$\qquad$_____ %$\qquad$_____ %

Average percentage of $NaHCO_3$$\qquad$_____ %

NAME _____

SECTION _____

POSTLABORATORY ASSIGNMENT

1. A 1.500-g sample of potassium hydrogen carbonate is decomposed by heating to produce 1.040 g of potassium carbonate. Calculate the theoretical yield and percent yield of K_2CO_3.

$$2\ KHCO_3(s) \xrightarrow{\Delta} K_2CO_3(s)\ +\ H_2O(g)\ +\ CO_2(g)$$

_____

2. A 1.750-g sample containing potassium hydrogen carbonate is decomposed by heating. If the mass loss is 0.271 g, what is the percentage of $KHCO_3$ in the unknown mixture?

$$2\ KHCO_3(s) \xrightarrow{\Delta} K_2CO_3(s)\ +\ H_2CO_3(g)$$

_____

3. Solid potassium nitrate is decomposed by heating to give potassium nitrite and oxygen gas. If 1.00 g of potassium nitrate is heated, what is the milliliter volume of oxygen gas at STP?

$$2\ KNO_3(s) \xrightarrow{\Delta} 2\ KNO_2(s)\ +\ O_2(g)$$

_____

4. Potassium metal reacts with water to give potassium hydroxide and hydrogen gas. If 50.0 mL of hydrogen gas is produced at STP, what is the mass of potassium that reacted?

$$2\ K(s)\ +\ 2\ H_2O(l)\ \rightarrow\ 2\ KOH(aq)\ +\ H_2(g)$$

_____

5. (optional) The Solvay process is an industrial method for preparing baking soda. In the process, CO_2, NH_3, H_2O, and NaCl react to produce $NaHCO_3$. If 10.0 L of carbon dioxide gas at STP and 50.0 g of sodium chloride react with excess ammonia and water, what is the limiting reactant and the mass of baking soda produced?

$$CO_2(g)\ +\ NH_3(g)\ +\ H_2O(l)\ +\ NaCl(s)\ \rightarrow\ NaHCO_3(s)\ +\ NH_4Cl(aq)$$

_____

Precipitating Calcium Phosphate

- To determine the percent yield of calcium phosphate from a precipitation reaction.
- To determine the percentage of calcium chloride in an unknown mixture.
- To gain proficiency in transferring and filtering a precipitate.

In this experiment, we will use aqueous solutions to produce a precipitate of insoluble calcium phosphate, $Ca_3(PO_4)_2$. The equation for the reaction is

$$3\ CaCl_2(aq)\ +\ 2\ Na_3PO_4(aq)\ \rightarrow\ Ca_3(PO_4)_2(s)\ +\ 6\ NaCl(aq)$$

The precipitate will be collected in filter paper, which separates the insoluble particles from aqueous solution. The experimental mass of the precipitate is referred to as the **actual yield**. The calculated mass of the precipitate using the above equation is referred to as the **theoretical yield**.

The **percent yield** from a chemical reaction is an expression for the amount of actual yield compared to the theoretical yield. While some experimental errors lead to high results, other errors may give low results. Thus, the percent yield can be greater than—or less than—100%.

Percent Yield of Calcium Phosphate from Calcium Chloride

Example Exercise 15.1   •   % Yield of $Ca_3(PO_4)_2$ from $CaCl_2$

A 0.555-g sample of calcium chloride gives a 0.525-g precipitate of calcium phosphate. Calculate the theoretical yield and percent yield of $Ca_3(PO_4)_2$.

Solution: According to the balanced chemical equation, 3 mol $CaCl_2$ (110.98 g/mol) produce 1 mol $Ca_3(PO_4)_2$ (310.18 g/mol) precipitate. We can calculate the theoretical mass of precipitate as follows:

$$0.555 \text{ g } CaCl_2 \times \frac{1 \text{ mol } CaCl_2}{110.98 \text{ g } CaCl_2} \times \frac{1 \text{ mol } Ca_3(PO_4)_2}{3 \text{ mol } CaCl_2} \times \frac{310.18 \text{ g } Ca_3(PO_4)_2}{1 \text{ mol } Ca_3(PO_4)_2}$$

$$= \quad 0.517 \text{ g } Ca_3(PO_4)_2$$

Since the actual yield of $Ca_3(PO_4)_2$ is 0.525 g, the percent yield is

$$\frac{\text{actual yield}}{\text{theoretical yield}} \quad \times \quad 100\% \quad = \quad \% \text{ yield}$$

$$\frac{0.525 \text{ g}}{0.517 \text{ g}} \quad \times \quad 100\% \quad = \quad 102\%$$

Percentage of Calcium Chloride in an Unknown Mixture

When an unknown mixture containing calcium chloride reacts with sodium phosphate, the equation for the reaction is the same as above. In this calculation, however, we will relate the mass of calcium phosphate to the mass of calcium chloride in the original unknown mixture.

Example Exercise 15.2   •   % $CaCl_2$ in an Unknown Mixture

A 1.000-g unknown mixture containing calcium chloride gives a 0.565-g precipitate of calcium phosphate. Calculate the percentage of $CaCl_2$ in the unknown mixture.

Solution: In this example, we must relate the mass of precipitate product to the mass of the original $CaCl_2$ reactant.

$$0.565 \text{ g } Ca_3(PO_4)_2 \times \frac{1 \text{ mol } Ca_3(PO_4)_2}{310.18 \text{ g } Ca_3(PO_4)_2} \times \frac{3 \text{ mol } CaCl_2}{1 \text{ mol } Ca_3(PO_4)_2} \times \frac{110.98 \text{ g } CaCl_2}{1 \text{ mol } CaCl_2}$$

$$= \quad 0.606 \text{ g } CaCl_2$$

If the sample mixture has a mass of 1.000 g, the percentage of $CaCl_2$ is

$$\frac{\text{mass } CaCl_2}{\text{mass sample}} \quad \times \quad 100\% \quad = \quad \% \ CaCl_2$$

$$\frac{0.606 \text{ g}}{1.000 \text{ g}} \quad \times \quad 100\% \quad = \quad 60.6\%$$

- 250-mL beaker
- 100-mL graduated cylinder
- ring stand
- wire gauze
- clay triangle
- 400-mL beaker
- wash bottle with distilled water
- glass stirring rod + rubber policeman

- calcium chloride, anhydrous $CaCl_2$
- sodium phosphate solution, 0.5 M Na_3PO_4
- filter paper
- long-stem funnel (75 mm diameter)
- unknown calcium chloride mixtures (50–70% $CaCl_2$)

PROCEDURE

A. Percent Yield of $Ca_3(PO_4)_2$ from $CaCl_2$

1. Place a 250-mL beaker on the balance, and record the mass. Add about 0.5 g of calcium chloride, and reweigh accurately.

2. Dissolve the $CaCl_2$ sample completely in 50 mL of distilled water. Using a graduated cylinder, add 10 mL of 0.5 M Na_3PO_4 solution to the sample in the 250-mL beaker.

3. Support the beaker with a wire gauze on a ring stand. Bring the solution to a gentle boil and then turn off the burner. Allow the precipitate to digest until the solution is cool.

   Note: As the precipitate settles from aqueous solution, add a few drops of Na_3PO_4 solution to test for completeness of precipitation. If the clear solution becomes cloudy, add more Na_3PO_4 solution to assure that all of the calcium has been precipitated from solution.

4. Weigh a disk of filter paper. Prepare a filter paper cone by folding the disk twice. Insert the filter paper into the funnel, and moisten with distilled water using the wash bottle.

5. Assemble a filtering apparatus as shown in Figure 15.1.

6. Without disturbing the precipitate, carefully pour off the supernate into the filter paper, using a stirring rod to guide the flow as shown in Figure 15.1(a). Rinse out the bulk of the precipitate with a stream of water from the wash bottle as shown in Figure 15.1(b). Clean the beaker using a rubber policeman and rinse the residue into the filter paper.

   Note: The precipitate is "jelly-like" and filters quite slowly. To speed filtration, first pour off the clear supernate. Then transfer the precipitate using a minimum amount of wash water. To avoid delay, allow the filtration to continue and begin the unknown part of the experiment.

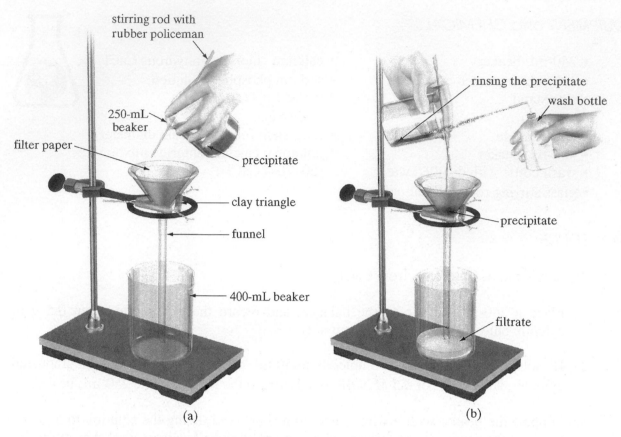

stirring rod with
rubber policeman

250-mL
beaker

filter paper

precipitate

clay triangle

funnel

400-mL beaker

(a)

rinsing the precipitate

wash bottle

precipitate

filtrate

(b)

Figure 15.1 Apparatus for Filtration (a) When the precipitate settles, pour off the supernate. (b) After the supernate passes through the filter paper, rinse the precipitate into the filter paper using a stream of water from a wash bottle.

7. After the supernate has passed through the filter, *carefully* remove the paper cone from the funnel. After the precipitate is completely dry, weigh the filter paper and precipitate.

   Note: If an oven is available, dry the precipitate overnight at ~110°C.

8. Calculate the theoretical yield of calcium phosphate from the mass of the calcium chloride. Find the percent yield.

B. Percentage of $CaCl_2$ in an Unknown Mixture

1. Obtain an unknown sample mixture containing calcium chloride, $CaCl_2$. Record the unknown number in the Data Table.

2. Place a 250-mL beaker on the balance and record the mass. Add about 1 g of unknown mixture, and reweigh accurately.

3. Repeat steps 1–7 as in Procedure A; substitute the unknown mixture for calcium chloride.

4. Calculate the mass of calcium chloride in the unknown sample from the mass of precipitate. Find the percentage of calcium chloride in the unknown mixture.

EXPERIMENT 15 NAME ____________________

DATE ____________________ SECTION ____________________

*PRELABORATORY ASSIGNMENT**

1. In your own words, define the following terms:

   actual yield

   digestion

   filtrate

   percent yield

   supernate

   theoretical yield

2. What difficulty arises if the precipitate is not allowed to settle completely from solution?

3. What is the purpose of the rubber policeman?

4. What should be done if particles of precipitate appear in the filtrate?

5. Is it possible to have a percent yield of calcium phosphate that is greater than 100%?

---

* *Answers in Appendix J*

6. A sample of calcium chloride reacts with sodium phosphate to give a precipitate of calcium phosphate. Refer to Example Exercise 15.1 and determine the theoretical yield of $Ca_3(PO_4)_2$, from the following data:

mass of beaker + $CaCl_2$	100.621 g
mass of beaker	100.111 g
mass of $Ca_3(PO_4)_2$ ppt	0.466 g

   What is the percent yield of calcium phosphate, $Ca_3(PO_4)_2$?

7. A 0.995-g unknown mixture containing calcium chloride reacts to give 0.505 g precipitate of calcium phosphate. Refer to Example Exercise 15.2 and find the mass of calcium chloride, $CaCl_2$, in the unknown mixture from the following data:

mass of beaker + unknown mixture	105.020 g
mass of beaker	104.025 g
mass of $Ca_3(PO_4)_2$ ppt	0.505 g

   What is the percentage of calcium chloride, $CaCl_2$, in the unknown mixture?

8. What are the major sources of error in this experiment?

9. What safety precautions must be observed in this experiment?

NAME _____

DATE _____

SECTION _____

DATA TABLE

A. Percent Yield of $Ca_3(PO_4)_2$ from $CaCl_2$

mass of beaker + $CaCl_2$ _____ g

mass of beaker _____ g

mass of $CaCl_2$ _____ g

mass of filter paper + $Ca_3(PO_4)_2$ ppt _____ g

mass of filter paper _____ g

mass of $Ca_3(PO_4)_2$ ppt *(actual yield)* _____ g

Show the calculation for theoretical yield of $Ca_3(PO_4)_2$ for trial 1 (see Example Exercise 15.1).

mass of $Ca_3(PO_4)_2$ ppt *(theoretical yield)* _____ g

Show the calculation for percent yield of $Ca_3(PO_4)_2$ for trial 1 (see Example Exercise 15.1).

Percent Yield of $Ca_3(PO_4)_2$ _____ %

B. Percentage of $CaCl_2$ in an Unknown Mixture   **UNKNOWN #** _____

      mass of beaker + unknown mixture   _____ g

      mass of beaker   _____ g

      mass of unknown mixture   _____ g

      mass of filter paper + $Ca_3(PO_4)_2$ ppt   _____ g

      mass of filter paper   _____ g

      mass of $Ca_3(PO_4)_2$ ppt   _____ g

Show the calculation for the mass of $CaCl_2$ in the unknown mixture for trial 1 (see Example Exercise 15.2).

      mass of $CaCl_2$ _____ g

Show the calculation for the percentage of $CaCl_2$ in the unknown mixture for trial 1 (see Example Exercise 15.2).

      Percentage of $CaCl_2$ _____ %

EXPERIMENT 15 NAME ____________________

DATE ____________________ SECTION ____________________

POSTLABORATORY ASSIGNMENT

1. A 1.000-g sample of stannous fluoride gives a 1.175-g precipitate of stannous phosphate. Calculate the theoretical yield and percent yield of $Sn_3(PO_4)_2$.

$$3\ SnF_2(aq) + 2\ K_3PO_4(aq) \rightarrow Sn_3(PO_4)_2(s) + 6\ KF(aq)$$

____

2. A 10.000-g sample of toothpaste containing stannous fluoride gives a 0.145-g precipitate of stannous phosphate. What is the percentage of SnF_2 in the toothpaste sample?

$$3\ SnF_2(aq) + 2\ K_3PO_4(aq) \rightarrow Sn_3(PO_4)_2(s) + 6\ KF(aq)$$

____

3. Potash is composed mainly of potassium carbonate, K_2CO_3. If 1.000 g of potash reacts with sulfuric acid to produce 145 mL of carbon dioxide gas at STP, what is the percentage of potassium carbonate in the marble sample?

$$K_2CO_3(s) \;+\; H_2SO_4(aq) \;\rightarrow\; K_2SO_4(aq) \;+\; H_2O(l) \;+\; CO_2(g)$$

_____

4. Tooth enamel is mainly hydroxyapatite, $Ca_{10}(PO_4)_6(OH)_2$. If 1.117 g of tooth enamel reacts with acid to give 1.111 g of calcium chloride, what is the percentage of hydroxyapatite in the tooth enamel sample?

$$Ca_{10}(PO_4)_6(OH)_2(s) \;+\; 20\,HCl(aq) \;\rightarrow\; 10\,CaCl_2(aq) \;+\; 6\,H_3PO_4(aq) \;+\; 2\,H_2O(l)$$

_____

5. (optional) Milk of magnesia, $Mg(OH)_2$, is prepared from aqueous solutions of magnesium sulfate and sodium hydroxide. If a solution containing 50.0 g of $MgSO_4$ is added to a solution with 50.0 g of NaOH, what is the limiting reactant and mass of milk of magnesia produced?

$$MgSO_4(aq) \;+\; 2\,NaOH(aq) \;\rightarrow\; Mg(OH)_2(s) \;+\; Na_2SO_4(aq)$$

_____

EXPERIMENT

16

Generating Hydrogen Gas

OBJECTIVES

- To determine the experimental molar volume of hydrogen gas at STP.
- To determine the atomic mass and identity for an unknown metal (**X**).
- To gain experience in collecting a gas over water and reading a barometer.

DISCUSSION

The **molar volume** of a gas is the volume occupied by one mole of gas at standard conditions. The theoretical value for the molar volume of an ideal gas at **standard temperature and pressure (STP)** is 22.4 liters. A volume of 22.4 L contains Avogadro's number of molecules. The molar volume concept is illustrated for hydrogen gas in Figure 16.1.

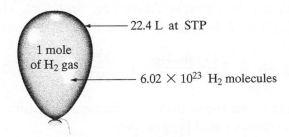

Figure 16.1 The Mole Concept One mole of hydrogen gas occupies 22.4 L at STP, and contains Avogadro's number of molecules.

Molar Volume of Hydrogen Gas

In this experiment, magnesium metal reacts with hydrochloric acid according to the equation

$$Mg(s) \ + \ 2\,HCl(aq) \ \rightarrow \ MgCl_2(aq) \ + \ H_2(g)$$

We can find an experimental value for the molar volume of hydrogen gas from the stoichiometry of the reaction. The following example exercise illustrates the calculation of molar volume.

Example Exercise 16.1 • Molar Volume of Hydrogen Gas

A 0.0750 g sample of magnesium metal reacts with hydrochloric acid to produce 77.5 mL of hydrogen gas. The "wet" gas is collected over water at 20°C and an atmospheric pressure of 763 mm Hg. Calculate the experimental molar volume of hydrogen gas at STP.

Solution: Since the gas is collected over water, both hydrogen gas and water vapor contribute to the total pressure. We find in Table 16.1 the **vapor pressure** of water at 20°C is 18 mm Hg. The total pressure of the hydrogen gas and water vapor equals the **atmospheric pressure**, that is, 763 mm Hg. Applying **Dalton's law of partial pressures** we have

$$
\begin{aligned}
P_{H_2} + P_{H_2O} &= P_{atm} \\
P_{H_2} + 18 \text{ mm Hg} &= 763 \text{ mm Hg} \\
P_{H_2} &= 763 \text{ mm Hg} - 18 \text{ mm Hg} \\
&= 745 \text{ mm Hg}
\end{aligned}
$$

Let's prepare a table for the pressure, volume, and temperature data.

Conditions	P	V	T
initial	745 mm Hg	77.5 mL	20 + 273 = 293 K
final	760 mm Hg	V_{STP}	0 + 273 = 273 K

We can correct the volume of H_2 gas to STP using the **combined gas law**.

$$V_{initial} \ \times \ P_{factor} \ \times \ T_{factor} \ = \ V_{STP}$$

The pressure increases, so the volume decreases. The P_{factor} is less than 1. The temperature decreases, so the volume decreases. The T_{factor} is less than 1.

$$77.5 \text{ mL } H_2 \times \frac{745 \text{ mm Hg}}{760 \text{ mm Hg}} \times \frac{273 \text{ K}}{293 \text{ K}} \ = \ 70.8 \text{ mL } H_2$$

Referring to the above balanced chemical equation, we see that 1 mol Mg metal produces 1 mol H_2 gas; thus,

$$0.0750 \text{ g Mg} \times \frac{1 \text{ mol Mg}}{24.31 \text{ g Mg}} \times \frac{1 \text{ mol } H_2}{1 \text{ mol Mg}} \ = \ 0.00309 \text{ mol } H_2$$

The molar volume is the ratio of liters of H_2 gas at STP to moles of H_2 gas produced from the reaction. Thus,

$$\frac{70.8 \; \text{mL } H_2}{0.00309 \; \text{mol } H_2} \; \times \; \frac{1 \; L}{1000 \; \text{mL}} \; = \; 22.9 \; \text{L/mol}$$

The experimental molar volume is 22.9 L/mol at STP. This value compares closely with the theoretical molar volume of 22.4 L/mol.

Atomic Mass of an Unknown Metal

After reacting magnesium metal and hydrochloric acid, we will react an unknown metal (**X**) in a similar fashion. The equation for the reaction is

$$\mathbf{X}(s) \; + \; 2 \, HCl(aq) \; \rightarrow \; \mathbf{X}Cl_2(aq) \; + \; H_2(g)$$

We can use the concept of molar volume to calculate the atomic mass of the unknown metal as shown in the following example exercise.

Example Exercise 16.2 • Atomic Mass of an Unknown Metal

A 0.215-g sample of unknown metal produces 81.0 mL of hydrogen gas. The "wet" gas is collected over water at 21°C and an atmospheric pressure of 763 mm Hg. Calculate the atomic mass of the unknown metal (**X**).

Solution: We find in Table 16.1 that the vapor pressure of water at 21°C is 19 mm Hg. The partial pressure of hydrogen gas is found by applying Dalton's law.

$$
\begin{aligned}
P_{H_2} + P_{H_2O} &= P_{atm} \\
P_{H_2} + 19 \; \text{mm Hg} &= 763 \; \text{mm Hg} \\
P_{H_2} &= 763 \; \text{mm Hg} - 19 \; \text{mm Hg} \\
&= 744 \; \text{mm Hg}
\end{aligned}
$$

Let's prepare a table for the pressure, volume, and temperature data.

Conditions	P	V	T
initial	744 mm Hg	81.0 mL	21 + 273 = 294 K
final	760 mm Hg	V_{STP}	0 + 273 = 273 K

We can correct the initial volume of H_2 gas to STP conditions as follows:

$$81.0 \; \text{mL } H_2 \; \times \; P_{factor} \; \times \; T_{factor} \; = \; V_{STP}$$

The pressure increases, so the volume decreases. The P_{factor} is less than 1. The temperature decreases, so the volume decreases. The T_{factor} is less than 1.

Since the pressure and temperature factors are less than 1, we have

$$81.0 \text{ mL H}_2 \quad \times \quad \frac{744 \text{ mm Hg}}{760 \text{ mm Hg}} \quad \times \quad \frac{273 \text{ K}}{294 \text{ K}} \quad = \quad 73.6 \text{ mL H}_2$$

According to the balanced equation for the reaction, 1 mole of unknown metal (X) produces 1 mole of hydrogen gas. Thus, at STP

$$73.6 \text{ mL H}_2 \quad \times \quad \frac{1 \text{ L}}{1000 \text{ mL}} \quad \times \quad \frac{1 \text{ mol H}_2}{22.4 \text{ L H}_2} \quad \times \quad \frac{1 \text{ mol X}}{1 \text{ mol H}_2} \quad = \quad 0.00329 \text{ mol X}$$

The atomic mass of the unknown metal is expressed by the ratio of the mass of sample to the moles of metal.

$$\frac{0.215 \text{ g X}}{0.00329 \text{ mol X}} \quad = \quad 65.3 \text{ g/mol}$$

The atomic mass of the unknown metal (X) is 65.3 g/mol. If we refer to the periodic table, we identify the unknown metal as zinc (65.39 g/mol).

In this experiment, we will collect hydrogen gas over water. The hydrogen gas displaces water from a graduated cylinder. Figure 16.2 illustrates an apparatus for collecting the gas.

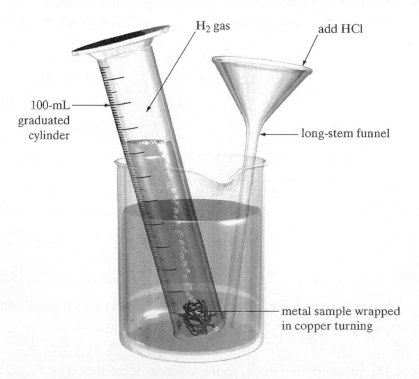

Figure 16.2 Apparatus for Gas Collection The volume of hydrogen gas produced equals the volume of water displaced from the graduated cylinder.

- 1000-mL beaker
- 100-mL graduated cylinder
- wash bottle with distilled water
- long-stem funnel
- 100-mL beaker
- 110°C thermometer
- barometer
- milligram balances
- tenth milligram balances (optional)

- magnesium, Mg ribbon
- copper, Cu light turnings
- dilute hydrochloric acid, 6 *M* HCl
- unknown metal samples (**X**)

PROCEDURE

A. Molar Volume of Hydrogen Gas

1. Cut a strip of magnesium metal ribbon having a mass between 0.070 and 0.090 g. (This mass corresponds to an approximate length of 7–9 cm of magnesium ribbon.) Weigh the Mg metal accurately. Roll the Mg ribbon into a compact coil, and completely wrap the metal with strands of copper turnings.

2. Add 700 mL of water into a 1000-mL beaker. Drop the copper-wrapped magnesium metal into the water.

3. Fill a 100-mL graduated cylinder with water. Adjust the water level to the upper rim using water from a wash bottle. Place a small piece of paper towel over the entire rim, and allow it to absorb water. Invert the graduated cylinder over the sink. Carefully put the graduated cylinder into the beaker. As the piece of towel floats free, place the graduated cylinder over the copper-wrapped metal as shown in Figure 16.2.

   Note: It is advisable to practice inverting the graduated cylinder filled with water. If the graduated cylinder loses water upon inversion, the paper towel may be too large. On occasion, the spout in the graduated cylinder is too deep. Exchanging the graduated cylinder solves this problem.

4. Using a long-stem funnel, add 25 mL of dilute hydrochloric acid into the beaker. Gas bubbles are observed when the acid reacts with the magnesium sample. When the reaction is complete, tilt the cylinder vertically and read the bottom of the meniscus. Record the volume (± 0.5 mL) in the Data Table.

5. Place the thermometer into the beaker of water, and observe the temperature. Record the temperature of the hydrogen gas in the Data Table.

Note: Since the hydrogen gas is collected over water, the temperature of the hydrogen gas is the same as the temperature of the water in the beaker.

6. Read the barometer, and record the atmospheric pressure. Find the vapor pressure of water from Table 16.1.

7. Calculate the molar volume of hydrogen gas at STP.

Table 16.1 Vapor Pressure of Water

Temperature	Pressure	Temperature	Pressure	Temperature	Pressure
16°C	14 mm Hg	21°C	19 mm Hg	26°C	25 mm Hg
17°C	15 mm Hg	22°C	20 mm Hg	27°C	27 mm Hg
18°C	16 mm Hg	23°C	21 mm Hg	28°C	28 mm Hg
19°C	17 mm Hg	24°C	22 mm Hg	29°C	30 mm Hg
20°C	18 mm Hg	25°C	24 mm Hg	30°C	32 mm Hg

B. Atomic Mass of an Unknown Metal

1. Obtain a sample of unknown metal (**X**), and record the number in the Data Table.

2. Follow the same steps as in Procedure A.

3. Calculate the atomic mass of the unknown metal (**X**). Refer to the atomic masses in the periodic table, and identify the unknown metal (**X**).

NAME _____

DATE _____   SECTION _____

PRELABORATORY ASSIGNMENT*

1. In your own words, define the following terms:

    atmospheric pressure

    combined gas law

    Dalton's law of partial pressures

    molar volume

    standard temperature and pressure (STP)

    vapor pressure

2. Why must the mass of magnesium metal be no greater than 0.09 g?

3. How can you tell when the magnesium metal has reacted completely?

4. Explain the meaning of the following terms:

    (a) "wet" gas

    (b) "dry" gas

5. Hydrogen gas is collected over water at 23°C and an atmospheric pressure of 756 mm Hg. Refer to Table 16.1 and calculate the partial pressure of hydrogen gas.

_____

Answers in Appendix J

6. A 0.0885-g sample of magnesium metal reacts with hydrochloric acid to give 90.5 mL of hydrogen gas at 23°C and 756 mm Hg. Refer to Example Exercise 16.1 and calculate the volume of hydrogen gas at STP.

Refer to Example Exercise 16.1 and find the moles of hydrogen gas at STP.

What is the experimental molar volume of hydrogen gas?

7. What are the major sources of error in this experiment?

8. What safety precautions should be observed in this experiment?

DATA TABLE

A. Molar Volume of Hydrogen Gas

mass of magnesium	_____ g		_____ g	
volume of hydrogen gas	_____ mL		_____ mL	
temperature of hydrogen gas	_____ °C		_____ °C	
atmospheric pressure (see barometer)	_____ mm Hg		_____ mm Hg	
vapor pressure of water (see Table 16.1)	_____ mm Hg		_____ mm Hg	
partial pressure of hydrogen gas	_____ mm Hg		_____ mm Hg	

Correct the volume of hydrogen gas to STP for trial 1 (see Example Exercise 16.1).

volume of hydrogen gas (STP)	_____ mL	_____ mL

Show the calculation for the moles of hydrogen gas for trial 1 (see Example Exercise 16.1).

moles of hydrogen gas	_____ mol	_____ mol

Show the calculation for the molar volume of hydrogen gas at STP for trial 1.

Molar volume of hydrogen gas (STP)	_____ L/mol	_____ L/mol

B. Atomic Mass of an Unknown Metal   **UNKNOWN #** _____

mass of metal (**X**)   _____ g   _____ g

volume of hydrogen gas   _____ mL   _____ mL

temperature of hydrogen gas   _____ °C   _____ °C

atmospheric pressure (see barometer)   _____ mm Hg   _____ mm Hg

vapor pressure of water (see Table 16.1)   _____ mm Hg   _____ mm Hg

partial pressure of hydrogen gas   _____ mm Hg   _____ mm Hg

Correct the volume of hydrogen gas to STP for trial 1 (see Example Exercise 16.2).

volume of hydrogen gas (STP)   _____ mL   _____ mL

Show the calculation for moles of unknown metal (**X**) for trial 1 (see Example Exercise 16.2).

moles of unknown metal   _____ mol   _____ mol

Show the calculation for the atomic mass of the unknown metal (**X**) for trial 1.

Atomic mass of the unknown metal (**X**)   _____ g/mol _____ g/mol

Identity of the unknown metal (**X**)   _____   _____

NAME _____

DATE _____

SECTION _____

POSTLABORATORY ASSIGNMENT

1. A 0.215-g sample of zinc metal reacted with hydrochloric acid according to the following balanced chemical equation:

$$Zn(s) \quad + \quad 2\,HCl(aq) \quad \rightarrow \quad ZnCl_2(aq) \quad + \quad H_2(g)$$

The volume of hydrogen collected over water was 80.0 mL at 20°C and a barometer reading of 756 mm Hg. Calculate the STP molar volume for hydrogen.

_____

2. A 0.136-g sample of an unknown metal (**X**) reacted with hydrochloric acid according to the following balanced chemical equation:

$$2\,\mathbf{X}(s) \quad + \quad 6\,HCl(aq) \quad \rightarrow \quad 2\,\mathbf{X}Cl_3(aq) \quad + \quad 3\,H_2(g)$$

The volume of gas collected over water was 90.5 mL at 20°C and 756 mm Hg. Calculate the atomic mass of the unknown metal and refer to the periodic table to identify the metal.

_____ g/mol

_____ (**X**)

3. Why does the water remain in the inverted glass?

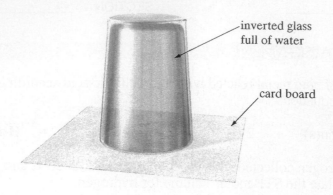

inverted glass
full of water

card board

4. A barometer reads 752 mm Hg. Express the atmospheric pressure in the following units.

(a) atm

(b) cm Hg

(c) psi

(d) in. Hg

5. (optional) A barometer in Denver, Colorado, reads 630 mm Hg. What is the height in feet of a barometer containing water? (*Hint:* Mercury is 13.6 times more dense than water.)

Generating Oxygen Gas

OBJECTIVES

- To determine the percentage of potassium chlorate in a known 90.0% mixture.
- To determine the percentage of potassium chlorate in an unknown mixture.
- To gain proficiency in decomposing a compound and collecting a gas over water.

DISCUSSION

When potassium chlorate, $KClO_3$, is heated, it decomposes to potassium chloride and oxygen gas. To assure a safe decomposition, manganese dioxide catalyst, MnO_2, must be mixed with the potassium chlorate. The equation for the decomposition of the black, powdery mixture is

$$2\ KClO_3(s) \quad \overset{MnO_2}{\underset{\Delta}{\rightarrow}} \quad 2\ KCl(s) \ + \ 3\ O_2(g)$$

We can determine the volume of oxygen gas produced by measuring the volume of water displaced. This technique is called **volume by displacement**. After correcting the volume of oxygen gas to standard conditions, we can use stoichiometry to relate the volume of oxygen gas to the mass of potassium chlorate in the original sample.

The percentage of potassium chlorate in a sample is found by comparing the mass of $KClO_3$ to the mass of the sample. The following example exercises illustrate the calculation for the percentage potassium chlorate in known and unknown sample mixtures.

Example Exercise 17.1 • % KClO₃ in a Known Mixture

A 0.930-g sample of a 90.0% mixture of $KClO_3$ is decomposed by heating. If 255 mL of oxygen gas is collected over water at 23°C and an atmospheric pressure of 758 mm Hg, what is the percentage of potassium chlorate in the known sample?

Solution: Since the gas is collected over water, both oxygen gas and water vapor contribute to the total pressure. We find the **vapor pressure** of water from Table 17.1 at 23°C is 21 mm Hg. The total pressure of the oxygen gas and water vapor equals the **atmospheric pressure**, 758 mm Hg. Applying **Dalton's law of partial pressures**, we have

$$P_{O_2} + P_{H_2O} = P_{atm}$$
$$P_{O_2} + 21 \text{ mm Hg} = 758 \text{ mm Hg}$$
$$P_{O_2} = 758 \text{ mm Hg} - 21 \text{ mm Hg}$$
$$P_{O_2} = 737 \text{ mm Hg}$$

Let's prepare a table for the pressure, volume, and temperature data.

Conditions	P	V	T
initial	737 mm Hg	255 mL	23 + 273 = 296 K
final	760 mm Hg	V_{STP}	0 + 273 = 273 K

We can correct the volume of O_2 gas to STP using the **combined gas law**.

$$V_{initial} \times P_{factor} \times T_{factor} = V_{STP}$$

The pressure increases, so the volume decreases. The P_{factor} is less than 1. The temperature decreases, so the volume decreases. The T_{factor} is less than 1.

$$255 \text{ mL } O_2 \times \frac{737 \text{ mm Hg}}{760 \text{ mm Hg}} \times \frac{273 \text{ K}}{296 \text{ K}} = 228 \text{ mL } O_2$$

Referring to the balanced chemical equation, we can see that 2 mol $KClO_3$ (122.55 g/mol) produces 3 mol O_2 gas. If we express **molar volume** as 22,400 mL, we can calculate the mass of $KClO_3$ in the 90.0% mixture.

$$228 \text{ mL } O_2 \times \frac{1 \text{ mol } O_2}{22,400 \text{ mL } O_2} \times \frac{2 \text{ mol } KClO_3}{3 \text{ mol } O_2} \times \frac{122.55 \text{ g } KClO_3}{1 \text{ mol } KClO_3} = 0.832 \text{ g } KClO_3$$

The percentage of $KClO_3$ in the original 0.930 g sample mixture is

$$\frac{\text{mass } KClO_3}{\text{mass sample}} \times 100\% = \% \text{ KClO}_3$$

$$\frac{0.832 \text{ g}}{0.930 \text{ g}} \times 100\% = 89.4\%$$

Since the result, 89.4%, agrees with the theoretical value, 90.0%, we can conclude that the experimental error is negligible.

Percentage of Potassium Chlorate in an Unknown Mixture

When an unknown mixture containing potassium chlorate is decomposed, the equation for the reaction is the same as above.

Example Exercise 17.2 • % KClO₃ in an Unknown Mixture

A 1.145-g sample of an unknown mixture is decomposed by heating. If 205 mL of oxygen gas is collected over water at 22°C and an atmospheric pressure of 758 mm Hg, what is the percentage of $KClO_3$ in the unknown sample?

Solution: We find in Table 17.1 that the vapor pressure of water at 22°C is 20 mm Hg. The partial pressure of oxygen gas is found by applying Dalton's law.

$$P_{O_2} + P_{H_2O} = P_{atm}$$
$$P_{O_2} + 20 \text{ mm Hg} = 758 \text{ mm Hg}$$
$$P_{O_2} = 758 \text{ mm Hg} - 20 \text{ mm Hg}$$
$$P_{O_2} = 738 \text{ mm Hg}$$

Let's prepare a table for the pressure, volume, and temperature data.

Conditions	P	V	T
initial	738 mm Hg	205 mL	22 + 273 = 295 K
final	760 mm Hg	V_{STP}	0 + 273 = 273 K

We can correct the initial volume of O_2 gas to STP conditions as follows:

$$V_{initial} \times P_{factor} \times T_{factor} = V_{STP}$$

The pressure increases, so the volume decreases. The P_{factor} is less than 1. The temperature decreases, so the volume decreases. The T_{factor} is less than 1.

$$205 \text{ mL } O_2 \times \frac{738 \text{ mm Hg}}{760 \text{ mm Hg}} \times \frac{273 \text{ K}}{295 \text{ K}} = 184 \text{ mL } O_2$$

According to the balanced equation, we see that 2 mol $KClO_3$ (122.55 g/mol) produces 3 mol O_2 gas. If we express the molar volume as 22,400 mL, we can calculate the mass of $KClO_3$ in the sample as follows:

$$184 \text{ mL } O_2 \times \frac{1 \text{ mol } O_2}{22,400 \text{ mL } O_2} \times \frac{2 \text{ mol } KClO_3}{3 \text{ mol } O_2} \times \frac{122.55 \text{ g } KClO_3}{1 \text{ mol } KClO_3} = 0.672 \text{ g } KClO_3$$

The percentage of $KClO_3$ in the original 1.145 g sample mixture is

$$\frac{\text{mass } KClO_3}{\text{mass sample}} \times 100\% = \% \ KClO_3$$

$$\frac{0.672 \text{ g}}{1.145 \text{ g}} \times 100\% = 58.7\%$$

Figure 17.1 shows the experimental apparatus for collecting oxygen gas. As the compound decomposes, oxygen gas is produced that displaces water from the Florence flask into a beaker. When the decomposition is complete, no more gas is released and the water level in the beaker remains constant. After a few minutes, the water level may actually decrease owing to the cooling of the oxygen gas in the Florence flask.

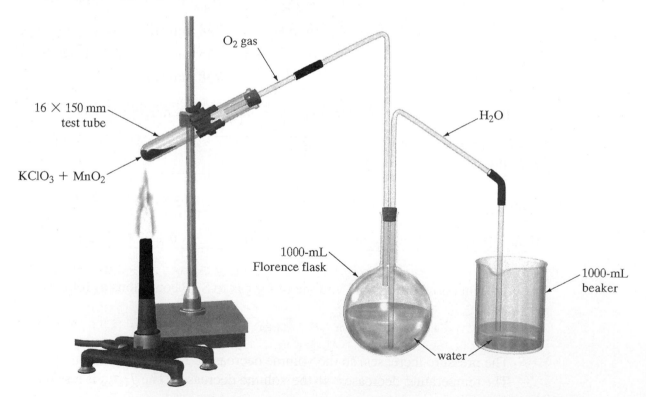

O_2 gas

16 × 150 mm test tube

H_2O

$KClO_3 + MnO_2$

1000-mL Florence flask

1000-mL beaker

water

Figure 17.1 Apparatus for Decomposition When the water level inside the beaker remains constant, the decomposition of $KClO_3$ is complete.

- 16 × 150 mm test tube
- 1000-mL Florence flask
- 1000-mL beaker
- 110°C thermometer
- barometer

- gas collection apparatus (see Figure 17.1)
- known potassium chlorate mixture, 90.0% $KClO_3$
- unknown potassium chlorate mixture, 40–80% $KClO_3$

PROCEDURE

A. Percentage of $KClO_3$ in a Known Mixture

1. Weigh a 16 × 150 mm test tube and record the mass. Add about 1 g of the known 90.0% potassium chlorate mixture, and reweigh accurately.

2. Set up the apparatus as shown in Figure 17.1. Fill the Florence flask to the neck with tap water, and insert the gas collection apparatus. Insert the small rubber stopper into the test tube as shown.

   Caution: Do not let any of the mixture contact the rubber stopper in the test tube. Ask the Instructor to inspect the apparatus before continuing.

3. Begin heating the test tube gently. As the water level in the beaker increases, heat the test tube more strongly. After the water level remains constant for a couple of minutes, discontinue heating and allow the test tube to cool for 10 minutes.

4. Weigh the 1000-mL beaker containing water on a platform balance (see **Appendix B**), and record the mass. Pour out the water and reweigh the empty beaker on the platform balance. Find the mass of water by difference. Record the volume of water and the volume of O_2 gas in the Data Table.

   Note: Since the density of water is 1.00 g/mL, the mass of water in grams is equal to its volume in milliliters. We assume that the volume of water displaced is equal to the volume of O_2 gas produced from the reaction.

5. Place the thermometer in the water in the Florence flask, and record the temperature of the O_2 gas in the Data Table.

   Note: Since the O_2 gas is collected over water, the temperature of the oxygen gas is the same as the temperature of the water in the flask.

6. Read the barometer, and record the atmospheric pressure. Find the vapor pressure of water from Table 17.1.

7. Correct the volume of oxygen gas to STP. Calculate the mass of potassium chlorate that decomposed and the percentage of $KClO_3$ in the mixture.

Table 17.1 Vapor Pressure of Water

Temperature	Pressure	Temperature	Pressure	Temperature	Pressure
16°C	14 mm Hg	21°C	19 mm Hg	26°C	25 mm Hg
17°C	15 mm Hg	22°C	20 mm Hg	27°C	27 mm Hg
18°C	16 mm Hg	23°C	21 mm Hg	28°C	28 mm Hg
19°C	17 mm Hg	24°C	22 mm Hg	29°C	30 mm Hg
20°C	18 mm Hg	25°C	24 mm Hg	30°C	32 mm Hg

B. Percentage of $KClO_3$ in an Unknown Mixture

1. Obtain an unknown sample mixture containing potassium chlorate, $KClO_3$. Record the unknown number in the Data Table.
2. Repeat steps 1–7 as in Procedure A; substitute an unknown mixture for the 90.0% potassium chlorate mixture.

PRELABORATORY ASSIGNMENT*

1. In your own words, define the following terms:

    catalyst

    combined gas law

    Dalton's law of partial pressures

    molar volume

    standard temperature and pressure (STP)

    vapor pressure

    volume by displacement

2. Why is manganese dioxide added to potassium chlorate before heating?

3. How can you tell when the sample is completely decomposed?

4. How do you determine the experimental volume of oxygen gas produced by the reaction?

5. Oxygen gas is collected over water at 22°C and an atmospheric pressure of 765 mm Hg. Refer to Table 17.1 and calculate the partial pressure of oxygen gas.

_____

* Answers in Appendix J

6. A 1.015-g sample of the known mixture of $KClO_3$ is decomposed by heating to give 278 mL of oxygen gas collected over water at 22°C and a barometer reading of 765 mm Hg. Refer to Example Exercise 17.1 and calculate the volume of oxygen gas at STP.

Refer to Example Exercise 17.1 and find the mass of $KClO_3$ (122.55 g/mol) in the mixture.

What is the percentage of potassium chlorate in the known mixture?

7. What are the major sources of error in this experiment?

8. What safety precautions must be observed in this experiment?

NAME _____

SECTION _____

DATA TABLE

A. Percentage of $KClO_3$ in a Known Mixture

mass of test tube + $KClO_3$ mixture	_____ g	_____ g
mass of test tube	_____ g	_____ g
mass of $KClO_3$ mixture	_____ g	_____ g
mass of beaker + water (platform balance)	_____ g	_____ g
mass of beaker (platform balance)	_____ g	_____ g
mass of water	_____ g	_____ g
volume of water	_____ mL	_____ mL
volume of O_2 gas	_____ mL	_____ mL
temperature of O_2 gas	_____ °C	_____ °C
atmospheric pressure (see barometer)	_____ mm Hg	_____ mm Hg
vapor pressure of water (see Table 17.1)	_____ mm Hg	_____ mm Hg
partial pressure of O_2 gas	_____ mm Hg	_____ mm Hg

Show the calculation for the volume of O_2 gas at STP for trial 1 (see Example Exercise 17.1).

    volume of O_2 gas (STP)   _____ mL   _____ mL

Show the calculation for the percentage of $KClO_3$ in the known mixture for trial 1.

    Percentage of $KClO_3$ _____ %   _____ %

B. Percentage of $KClO_3$ in an Unknown Mixture   **UNKNOWN #** _____

mass of test tube + $KClO_3$ mixture   _____ g   _____ g

mass of test tube   _____ g   _____ g

mass of $KClO_3$ mixture   _____ g   _____ g

mass of beaker + water (platform balance)   _____ g   _____ g

mass of beaker (platform balance)   _____ g   _____ g

mass of water   _____ g   _____ g

volume of water   _____ mL   _____ mL

volume of O_2 gas   _____ mL   _____ mL

temperature of O_2 gas   _____ °C   _____ °C

atmospheric pressure (see barometer)   _____ mm Hg   _____ mm Hg

vapor pressure of water (see Table 17.1)   _____ mm Hg   _____ mm Hg

partial pressure of O_2 gas   _____ mm Hg   _____ mm Hg

Show the calculation for the volume of O_2 gas at STP for trial 1 (see Example Exercise 17.2).

volume of O_2 gas (STP)   _____ mL   _____ mL

Show the calculation for the percentage of $KClO_3$ in the unknown mixture for trial 1.

Percentage of $KClO_3$ _____ %   _____ %

Average percentage of $KClO_3$ _____ %

POSTLABORATORY ASSIGNMENT

1. A 1.000-g sample of a 90.0% lithium chlorate mixture is decomposed by heating. If 370.0 mL of oxygen gas is collected over water at 20°C and an atmospheric pressure of 756 mm Hg, what is the experimental percentage of lithium chlorate in the mixture?

$$2 \ LiClO_3(s) \quad \overset{\Delta}{\rightarrow} \quad 2 \ LiCl(s) \ + \ 3 \ O_2(g)$$

_____

2. A 1.500-g sample of a lithium chlorate mixture is decomposed by heating. If 308.0 mL of oxygen gas is collected over water at 20°C and 756 mm Hg, what is the percentage of lithium chlorate in the mixture?

$$2 \ LiClO_3(s) \quad \overset{\Delta}{\rightarrow} \quad 2 \ LiCl(s) \ + \ 3 \ O_2(g)$$

_____

3. If a 0.500-g sample of a potassium nitrate mixture is decomposed by heating, what is the milliliter volume of oxygen gas released at STP?

$$2 \, KNO_3(s) \quad \xrightarrow{\Delta} \quad 2 \, KNO_2(s) \; + \; O_2(g)$$

_____

4. A sample of mercury(II) oxide upon heating released oxygen gas. If 50.0 mL of oxygen gas is produced at STP, what is the mass of mercuric oxide that decomposed?

$$2 \, HgO(s) \quad \xrightarrow{\Delta} \quad 2 \, Hg(l) \; + \; O_2(g)$$

_____

5. (optional) An industrial process for manufacturing potassium chlorate involves passing chlorine gas through hot potassium hydroxide solution. If 25.0 L of chlorine gas at STP is passed through an aqueous solution containing 150.0 g of potassium hydroxide, what is the limiting reactant and the mass of potassium chlorate produced?

$$3 \, Cl_2(g) \; + \; 6 \, KOH(aq) \quad \rightarrow \quad KClO_3(s) \; + \; 5 \, KCl(aq) \; + \; 3 \, H_2O(l)$$

_____

Molecular Models and Chemical Bonds

- To construct models of molecules with single, double, and triple bonds.
- To draw the structural formula for a molecule based on the molecular model.
- To draw the electron dot formula corresponding to the structural formula.
- To draw the structural and electron dot formulas for unknown molecular models.

DISCUSSION

The attraction between two atoms in a molecule is called a chemical bond. In a **covalent bond**, two nonmetal atoms are attracted to each other by sharing valence electrons. The **valence electrons** are the electrons farthest from the nucleus and occupy the highest s and p sublevels. The number of valence electrons is found from the periodic table. The group number of an element indicates the number of valence electrons. For example, fluorine is in Group VIIA/17 and has seven valence electrons (7 e$^-$).

Example Exercise 18.1 • Valence Electrons and the Periodic Table

Refer to the group number in the periodic table and determine the valence electrons for the following elements: (a) H; (b) C; and (c) O.

Solution:　(a)　Hydrogen is in group IA/1; thus, H has one valence electron.
　　　　　　(b)　Carbon is in Group IVA/14; thus, C has four valence electrons.
　　　　　　(c)　Chlorine is in Group VIIA/17; thus, Cl has seven valence electrons.

In this experiment, we will draw the **structural formula** and **electron dot formula** for molecules after building a model. A model is constructed from spherical balls and connectors, where each ball represents an atom and each connector a single bond. Since a **single bond** shares two electrons, each connector represents an electron pair.

A **double bond** shares two pairs of electrons. A molecular model is constructed using two connectors to represent the double bond. A **triple bond** shares three pairs of electrons. A molecular model is constructed using three connectors to represent the triple bond.

The following example exercises illustrate the structural formula and electron dot formula for molecular models having single, double, and triple bonds.

Example Exercise 18.2 • Structural and Electron Dot Formula for CHCl₃

The molecular model of chloroform is sketched below. Draw (a) the structural formula and (b) the electron dot formula. Each atom (excluding H) should be surrounded by an octet of electrons. (c) Verify the electron dot formula by checking the total number of electron dots against the sum of all valence electrons.

chloroform, CHCl₃

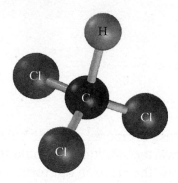

Solution: (a) Each stick represents a single bond, so the structural formula is

$$\begin{array}{c} H \\ | \\ Cl-C-Cl \\ | \\ Cl \end{array}$$

(b) Each dash in the structural formula indicates an electron pair; therefore,

$$\begin{array}{c} H \\ .. \\ Cl:C:Cl \\ .. \\ Cl \end{array}$$

Hydrogen and carbon are complete as shown; two electrons and eight electrons, respectively. However, each chlorine also requires an octet, which we will complete as follows:

$$\begin{array}{c} H \\ .. \quad .. \\ :Cl:C:Cl: \\ .. \quad .. \\ :Cl: \\ .. \end{array}$$

(c) To verify the above electron dot formula, we will find the sum of all valence electrons.

$$
\begin{aligned}
1\ \text{H}\ (1 \times 1\ \text{e--}) &= 1\ \text{e--} \\
1\ \text{C}\ (1 \times 4\ \text{e--}) &= 4\ \text{e--} \\
3\ \text{Cl}\ (3 \times 7\ \text{e--}) &= \underline{21\ \text{e--}} \\
\textit{sum of valence electrons} &= 26\ \text{e--}
\end{aligned}
$$

There are 26 valence electrons, and 26 dots were used in the electron dot formula; thus, the formula is verified.

Example Exercise 18.3 • Structural and Electron Dot Formula for H₂CO

A molecular model of formaldehyde is sketched below. Draw the (a) structural formula and (b) electron dot formula. (c) Find the sum of all valence electrons to verify the electron dot formula.

formaldehyde, H$_2$CO

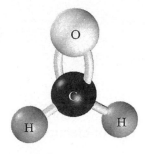

Solution: (a) Two connectors joining the carbon and oxygen atoms represent a double bond. The structural formula can be shown as

$$
\begin{array}{c}
\text{O} \\
\|\| \\
\text{H}-\text{C}-\text{H}
\end{array}
$$

(b) Each single bond contains one electron pair, and the double bond two electron pairs.

$$
\begin{array}{c}
\text{O} \\
\text{:\,:} \\
\text{H}:\text{C}:\text{H}
\end{array}
$$

Hydrogen shares two electrons and is stable. Carbon shares a total of eight electrons and satisfies the octet rule. Oxygen has only four of the eight electrons necessary to complete the octet. Therefore, we will add two unshared electron pairs.

$$
\begin{array}{c}
:\ddot{\text{O}} \\
\text{:\,:} \\
\text{H}:\text{C}:\text{H}
\end{array}
$$

(c) We can verify the above electron dot formula as follows:

$$2\ H\ (2 \times 1\ e-) = \quad 2\ e-$$
$$1\ C\ (1 \times 4\ e-) = \quad 4\ e-$$
$$1\ O\ (1 \times 6\ e-) = \quad \underline{6\ e-}$$
$$sum\ of\ valence\ electrons\ =\ 12\ e-$$

The 12 valence electrons equal the 12 electron dots and verify the formula.

Example Exercise 18.4 • Structural and Electron Dot Formula for HCN

A molecular model of hydrogen cyanide is sketched below. Draw (a) the structural formula and (b) the electron dot formula. (c) Verify the electron dot formula.

hydrogen cyanide, HCN

Solution: (a) The three connectors linking the carbon and nitrogen represent a triple pair of electrons.

$$H - C \equiv N$$

(b) We can draw an electron dot formula after realizing the triple bond contains three electron pairs.

$$H : C ::: N$$

In the above formula, nitrogen shares only six electrons. Therefore, we must add one unshared electron pair.

$$H : C ::: N :$$

(c) Let's verify the preceding electron dot formula.

$$1\ H\ (1 \times 1\ e-) = \quad 1\ e-$$
$$1\ C\ (1 \times 4\ e-) = \quad 4\ e-$$
$$1\ N\ (1 \times 5\ e-) = \quad \underline{5\ e-}$$
$$sum\ of\ valence\ electrons\ =\ 10\ e-$$

The 10 valence electrons verify the 10 e– dots.

- Molecular Model Kit
  Student molecular model sets (ISBN: 0-205-08136-3) are available
  from Prentice Hall @ 1-800-922-0579 (www.prenhall.com).

Directions for Using Molecular Models

When constructing a model, a hole in a ball represents a missing electron that is necessary to complete an octet. If two balls are joined by one connector, the connector represents a single bond composed of one electron pair. If two balls are joined by two connectors, the two connectors represent a double bond composed of two electron pairs. If two balls are joined by three connectors, the three connectors represent a triple bond composed of three electron pairs.

one rigid connector —	single bond
two flexible connectors —	double bond
three flexible connectors —	triple bond

A molecular model uses different color balls to represent hydrogen, carbon, oxygen, chlorine, bromine, iodine, and nitrogen atoms. The color code for each ball is as follows:

white ball —	hydrogen (one hole)
black ball —	carbon (four holes)
red ball —	oxygen (two holes)
green ball —	chlorine (one hole)
orange ball —	bromine (one hole)
purple ball —	iodine (one hole)
blue ball —	nitrogen (three holes)

Note: If the blue nitrogen ball has more than three holes, use a small peg or tape to fill the additional hole(s). All the holes in each ball must have a connector for a model to be built correctly.

PROCEDURE

1. Construct models for each of the molecules on the following page. Sketch the molecular model in the Data Table, showing its three-dimensional structure.

2. Draw the structural formula corresponding to the molecular model.

3. Draw the electron dot formula corresponding to the structural formula. Complete the octet by surrounding each atom with 8 electrons (2 electrons for a hydrogen atom).

4. Verify each electron dot formula by summing the valence electrons for the molecule, using the periodic table. This sum should equal the total number of dots in the electron dot formula.

A. Molecular Models with Single Bonds

(a) H_2 (b) Cl_2

(c) Br_2 (d) I_2

(e) HCl   (f) HBr

(g) ICl   (h) CH_4

(i) CH_2Cl_2 (j) HOCl

(k) H_2O_2 (l) NH_3

(m) N_2H_4 (n) NH_2OH

B. Molecular Models with Double Bonds

(a) O_2 (b) C_2H_4

(c) HONO   (d) HCOOH

(e) C_2H_3Cl

C. Molecular Models with Triple Bonds

(a) N_2 (b) C_2H_2

(c) HOCN

D. Molecular Models with Two Double Bonds

(a) CO_2 (b) C_3H_4

E. Unknown Molecular Models

The Instructor will provide models of unknown molecules. Draw the structural formula for each unknown model and the electron dot formula corresponding to each structural formula.

1. In your own words, define the following terms:

   covalent bond

   double bond

   electron dot formula

   octet rule

   single bond

   structural formula

   triple bond

   valence electrons

2. Refer to the periodic table in order to predict the number of valence electrons for each of the following elements:

   (a) oxygen

   (b) nitrogen

   (c) bromine

   (d) iodine

3. What do each of the following represent in the molecular model kit?

   (a) a white ball

   (b) a black ball

   (c) a red ball

   (d) a blue ball

   (e) one rigid connector

   (f) two flexible connectors

_____

* Answers in Appendix J

4. Draw the structural formula corresponding to each of the following molecular models.

(a)

(b)

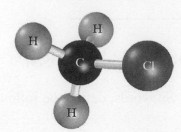

(c)

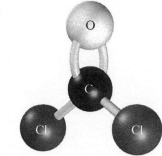

5. Draw the electron dot formula corresponding to each of the models in the preceding question.

(a)   (b)

(c)

6. Perform a valence electron check on each of the examples in the preceding questions.

   (a)   IBr   *Total Valence Electrons* ___

   (b) CH_3Cl *Total Valence Electrons* ___

   (c) Cl_2CO *Total Valence Electrons* ___

DATA TABLE

A. Molecular Models with Single Bonds   **Model Kit #_____**

Molecule	Molecular Model	Structural Formula	Electron Dot Formula	Valence Electron
(a) H_2				
(b) Cl_2				
(c) Br_2				
(d) I_2				

Molecule	Molecular Model	Structural Formula	Electron Dot Formula	Valence Electron
(e) HCl				
(f) HBr				
(g) ICl				
(h) CH$_4$				
(i) CH$_2$Cl$_2$				

Molecule	Molecular Model	Structural Formula	Electron Dot Formula	Valence Electron
(j) HOCl				
(k) H_2O_2				
(l) NH_3				
(m) N_2H_4				
(n) NH_2OH				

B. Molecular Models with Double Bonds

Molecule	Molecular Model	Structural Formula	Electron Dot Formula	Valence Electron
(a) O_2				
(b) C_2H_4				
(c) HONO				
(d) HCOOH				
(e) C_2H_3Cl				

C. Molecular Models with Triple Bonds

Molecule	Molecular Model	Structural Formula	Electron Dot Formula	Valence Electron
(a) N_2				
(b) C_2H_2				
(c) HOCN				

D. Molecular Models with Two Double Bonds

(a) CO_2				
(b) C_3H_4				

E. Unknown Molecular Models

Molecule	Molecular Model	Structural Formula	Electron Dot Formula	Valence Electron
#1				
#2				
#3				
#4				
#5				

1. Find the total number of valence electrons for the following molecules. Draw the electron dot formula and structural formula for each molecule. The central atom is in **bold.**

Molecule	Electron Dot Formula	Structural Formula
(a) $H_2\mathbf{S}$		
(b) $\mathbf{P}H_3$		
(c) $\mathbf{Si}H_4$		
(d) $\mathbf{C}S_2$		
(e) $\mathbf{S}O_2$		

2. Refer to the electron dot formula for molecules in the preceding question. Predict the electron pair geometry and the molecular shape for each molecule using VSEPR theory.

Molecule	Electron Pair Geometry	Molecular Shape
(a) H_2S		
(b) PH_3		
(c) SiH_4		
(d) CS_2		

3. Predict the bond angle in each of the following molecules based upon the molecular shape in the preceding question.

   (a) H_2S

   (b) PH_3

   (c) SiH_4

   (d) CS_2

4. (optional) Ultraviolet light from the Sun converts oxygen molecules in the upper atmosphere to ozone molecules. The UV light dissociates oxygen molecules into atoms, which can form a coordinate covalent bond to another oxygen molecule. Draw the electron dot formula and the structural formula for ozone, O_3.

Analysis of Saltwater

OBJECTIVES

- To observe the solubility of solid solutes in various solvents.
- To observe the miscibility of water and various solvents.
- To study the factors that affect the rate of dissolving.
- To demonstrate the behavior of a supersaturated solution.
- To determine the mass/mass percent concentration and molar concentration of sodium chloride in an unknown saltwater solution.
- To become proficient in pipetting and evaporating a solution to dryness.

DISCUSSION

The *like dissolves like* **rule** describes the general principle for dissolving a **solute** in a **solvent**. In general, ionic and polar solutes dissolve in polar solvents, and nonpolar solutes dissolve in nonpolar solvents. Water is a polar solvent and dissolves ionic compounds such as table salt, $NaCl$, and polar compounds such as table sugar, $C_{12}H_{22}O_{11}$. Hexane is a nonpolar solvent and does not dissolve salt or sugar. Hexane is a good solvent for nonpolar compounds such as grease and oil.

The general principle of *like dissolves like* also dictates whether two liquids are soluble. Liquids that dissolve in one another are said to be **miscible**. If both liquids are polar, they are miscible. Two liquids that are nonpolar also dissolve in one another. However, a polar liquid and a nonpolar liquid are **immiscible**. They are not soluble and separate into two layers.

The amount of solute dissolved in a solution can be expressed in many ways. For example, we can use the terms saturated and unsaturated to indicate the concentration of a solution. A *saturated solution* contains the maximum amount of solute that will dissolve at a given temperature. An *unsaturated solution* contains less than the maximum amount of solute. A **supersaturated solution** is unstable and contains more solute than ordinarily dissolves at a given temperature.

The **mass/mass percent concentration** (symbol **m/m %**) expresses the ratio of the mass of solute to the mass of solution. That is,

$$\frac{\text{mass of solute}}{\text{mass of solution}} \quad \times \quad 100\% \quad = \quad \text{m/m \%}$$

The following Example Exercise 19.1 illustrates a calculation to express the concentration of an aqueous saltwater solution as a mass/mass percent.

Example Exercise 19.1 • Mass/Mass Percent Concentration

A 10.0-mL sample of saltwater solution has a mass of 10.214 g. After evaporating to dryness, the solid NaCl residue weighs 0.305 g. Calculate the mass/mass percent concentration of the solution.

Solution: The mass/mass percent concentration is readily obtained by dividing the mass of NaCl solute by the mass of solution.

$$\frac{0.305 \text{ g NaCl}}{10.214 \text{ g solution}} \quad \times \quad 100\% \quad = \quad 2.99\% \text{ NaCl}$$

The **molar concentration** (symbol M) expresses the number of moles of solute in a liter of solution. Molar concentration is referred to as *molarity*, which we can show as follows.

$$\frac{\text{moles of solute}}{\text{liter of solution}} \quad = \quad M$$

The following Example Exercise 19.2 illustrates a calculation to express the concentration of an aqueous saltwater solution as molarity.

Example Exercise 19.2 • Molar Concentration

Calculate the molar concentration of sodium chloride in the saltwater solution evaporated to dryness in Example Exercise 19.1.

Solution: The molar mass of NaCl is 58.44 g/mol; therefore, the molarity is

$$\frac{0.305 \text{ g NaCl}}{10.0 \text{ mL solution}} \quad \times \quad \frac{1 \text{ mol NaCl}}{58.44 \text{ g NaCl}} \quad \times \quad \frac{1000 \text{ mL}}{1 \text{ L}} \quad = \quad \frac{0.522 \text{ mol NaCl}}{1 \text{ L solution}}$$

$$= \quad 0.522 \ M \text{ NaCl}$$

- 13 × 100 mm test tubes (6) and test tube rack
- 16 × 150 mm test tube
- thin glass stirring rod
- test tube brush
- wash bottle with distilled water
- mortar and pestle
- wire gauze
- evaporating dish
- 250-mL beaker
- 10-mL pipet and pipet bulb
- 100-mL beaker

- potassium permanganate, solid $KMnO_4$
- iodine, solid crystals I_2
- hexane, liquid C_6H_{14}
- methanol, liquid CH_3OH
- acetone, liquid C_3H_6O
- heptane, liquid C_7H_{16}
- ethanol, liquid C_2H_5OH
- rock salt, solid crystals $NaCl$
- sodium acetate trihydrate, solid crystals $NaC_2H_3O_2 \cdot 3\ H_2O$
- unknown saltwater solutions, 3.00–5.00% $NaCl$

PROCEDURE

A. Solutes and Solvent

1. *Solubility.* Place six *dry* test tubes in a test tube rack as shown in Figure 19.1. Add 10 drops of water to the first, 10 drops of hexane, C_6H_{14}, to the second, and 10 drops of methanol, CH_3OH, to the third. Drop a small crystal of potassium permanganate, $KMnO_4$, into each test tube. Record whether the crystal is *soluble* or *insoluble*.

   Repeat the procedure and add 10 drops of water to the first test tube, 10 drops of hexane to the second, and 10 drops of methanol to the third. Drop a small crystal of iodine, I_2, into each test tube. Record whether the crystal is *soluble* or *insoluble*.

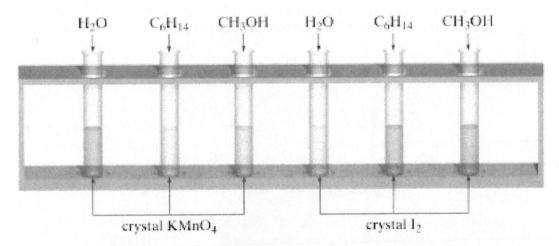

Figure 19.1 Solubility Crystals of $KMnO_4$ and I_2 are soluble in some solvents.

2. *Miscibility.* Put 10 drops of water into each of three test tubes shown in Figure 19.2. Add 10 drops of acetone, C_3H_6O, to the first test tube, 10 drops of heptane, C_7H_{16}, to the second, and 10 drops of ethanol, C_2H_5OH, to the third. Shake each test tube and record whether water and the solvent are *miscible* or *immiscible*.

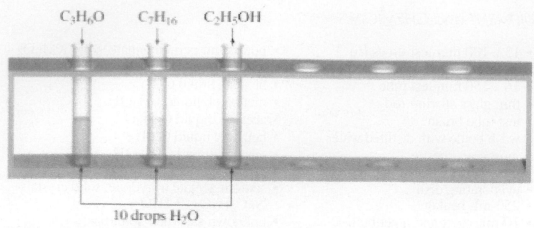

Figure 19.2 Miscibility Water is miscible with some organic solvents.

B. Rate of Dissolving

1. Half fill a test tube with distilled water. Add a single crystal of rock salt, and later record the length of time required for the crystal to dissolve at ~20 °C.

2. Set up the apparatus in Figure 19.3. Half fill three test tubes with distilled water, and place in a 250-mL beaker. Heat the water to a boil and shut off the burner.

3. Add a crystal of rock salt to test tube #1 in the beaker. Record the length of time required for the crystal to dissolve at ~100 °C.

4. Add a crystal of rock salt to test tube #2 in the beaker. Stir the solution, and record the length of time required for the crystal to dissolve at ~100 °C.

5. Grind a crystal of rock salt with mortar and pestle. Add the powder to test tube #3. Stir the solution, and record the time required for the powder to dissolve at ~100 °C.

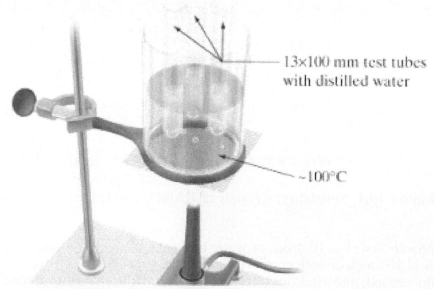

Figure 19.3 Boiling Water Apparatus The water in the beaker heats the solvent water in the test tubes to ~100°C.

C. Demonstration of Supersaturation

1. Set up the apparatus illustrated in Figure 19.4(a). Fill a 16 × 150 mm test tube 1/4 full with sodium acetate crystals, add enough distilled water to cover the crystals, and place the test tube in the utility clamp on the ring stand.

2. Heat the water in the beaker to a boil, and shut off the burner. Stir the solution in the test tube with a glass rod. After the crystals dissolve, move the test tube out of the beaker as shown in Figure 19.4(b).

3. Allow the solution to cool undisturbed to room temperature (~20°C). Drop one tiny seed crystal of sodium acetate trihydrate into the solution and observe the results.

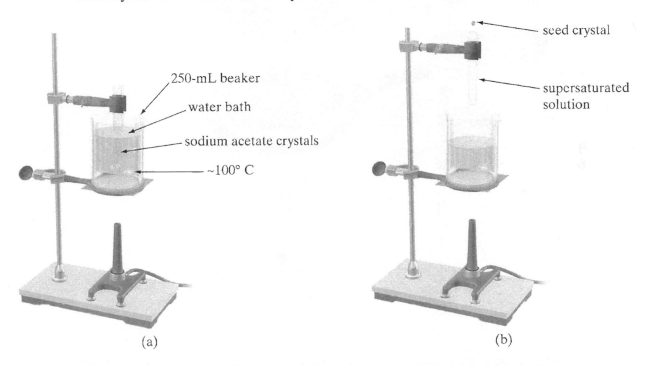

(a)   (b)

Figure 19.4 Supersaturation Apparatus (a) Sodium acetate crystals dissolve in water to give an unsaturated solution at ~100°C. (b) As the solution cools, it becomes saturated, and then supersaturated.

D. Concentration of Sodium Chloride in Saltwater

1. Obtain about 25 mL of unknown saltwater solution in a dry 100-mL beaker, and record the unknown number.

2. Weigh a dry evaporating dish on the balance.

3. Condition a pipet with the unknown saltwater solution, and transfer a 10.0-mL sample into the evaporating dish (see **Appendix E**). Reweigh the dish and solution.

4. Add about 200 mL of distilled water into a 250-mL beaker. Place the evaporating dish in the beaker and evaporate the solution to dryness (see Figure 19.5).

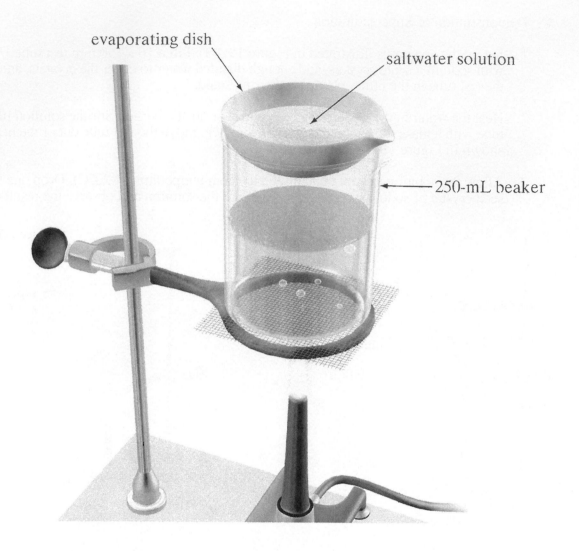
evaporating dish

saltwater solution

250-mL beaker

Figure 19.5 Evaporation of a Solution An unknown saltwater solution is
evaporated to dryness using a waterbath. As steam escapes from the beaker,
it may be necessary to add more distilled water to the beaker.

5. After evaporation, remove the dish and wipe the bottom of the dish dry. Hold the spout
of the dish with crucible tongs over a low flame to dry the last traces of moisture. Allow
the dish to cool, and weigh the evaporating dish with the solute residue.

Note: Do not heat the dish too strongly, as this may cause some of the residue to pop
from the dish.

6. Calculate the mass/mass percent and molar concentrations of sodium chloride in the
unknown saltwater solution.

PRELABORATORY ASSIGNMENT*

1. In your own words, define the following terms:

    conditioning

    immiscible

    like dissolves like rule

    mass/mass percent concentration (m/m %)

    miscible

    molar concentration (*M*)

    solute

    solvent

    supersaturated solution

2. How can you tell if a crystal is slightly soluble in a solvent?

3. Why is distilled water used in the waterbath?

4. Why is it necessary to stir the supersaturated solution after heating and before cooling? Why must you avoid jarring the test tube as it cools?

_____

Answers in Appendix J

5. A 10.0-mL sample of a saltwater solution containing sodium chloride, NaCl, was evaporated to dryness and gave the following data:

mass of evaporating dish + solution	50.827 g
mass of evaporating dish	40.505 g
mass of evaporating dish + NaCl	40.969 g

(a) Refer to Example Exercise 19.1 and calculate the mass/mass percent concentration of NaCl in the saltwater sample.

(b) Refer to Example Exercise 19.2 and find the molar concentration of NaCl (58.44 g/mol) in the saltwater sample.

6. What are the major sources of error in determining the concentration NaCl in saltwater?

7. What safety precautions must be observed in this experiment?

EXPERIMENT 19 NAME ____________________

DATE ____________________ SECTION ____________________

DATA TABLE

A. Solutes and Solvents

1. *Solubility*

SOLVENT

SOLUTE	*water* H_2O	*hexane* C_6H_{14}	*methanol* CH_3OH
$KMnO_4$			
I_2			

2. *Miscibility*

	acetone C_3H_6O	*heptane* C_7H_{16}	*ethanol* C_2H_5OH
H_2O			

B. Rate of Dissolving

Temperature	*Stirring*	*Particle Size*	*Time*
~20°C	No	Crystal	
~100°C	No	Crystal	
~100°C	Yes	Crystal	
~100°C	Yes	Powder	

C. Demonstration of Supersaturation

Observation

D. Concentration of Sodium Chloride in Saltwater   **UNKNOWN #** _____

      volume of saltwater solution   _____ mL   _____ mL

      mass of evaporating dish + solution   _____ g   _____ g
      (before heating)

      mass of evaporating dish   _____ g   _____ g

      mass of saltwater solution   _____ g   _____ g

      mass of evaporating dish + NaCl solute   _____ g   _____ g
      (after heating)

      mass of NaCl solute   _____ g   _____ g

Show the calculation for percent concentration of NaCl for trial 1 (see Example Exercise 19.1).

      Mass/mass percent concentration of NaCl   _____ %   _____ %

Show the calculation for molar concentration of NaCl for trial 1 (see Example Exercise 19.2).

      Molarity of NaCl   _____ *M*   _____ *M*

          Average molarity of NaCl   _____ *M*

POSTLABORATORY ASSIGNMENT

1. Indicate whether the solute solid is generally *soluble* or *insoluble* in each solvent liquid.

SOLVENT

SOLUTE	polar liquid	nonpolar liquid
ionic solid		
polar solid		
nonpolar solid		

2. Based on the "like dissolves like" rule, predict whether each of the following solutes is *soluble* or *insoluble* in water.

(a) ascorbic acid, $C_6H_8O_6$ _____

(b) cholesterol, $C_{27}H_{46}O$ _____

(c) fructose, $C_6H_{12}O_6$ _____

3. Indicate whether the solute liquid is generally *miscible* or *immiscible* in each solvent liquid.

SOLVENT

SOLUTE	polar liquid	nonpolar liquid
polar liquid		
nonpolar liquid		

4. Based on the "like dissolves like" rule, predict whether each of the following solvents is *miscible* or *immiscible* with water.

(a) octane, C_8H_{18} _____

(b) rubbing alcohol, C_3H_7OH _____

(c) toluene, $C_6H_5CH_3$ _____

5. Indicate whether the rate of dissolving *increases* or *decreases* for each of the following.

   (a) heating the solution

                                              _____

   (b) stirring the solution

                                              _____

   (c) grinding the solute

                                              _____

6. A 25.0-mL sample of potassium chloride solution was found to have a mass of 25.115 g. When the solution was heated to dryness, the mass of the solute residue was 0.252 g. Calculate the (a) mass/mass percent concentration and (b) molar concentration.

(a) _____

(b) _____

7. (optional) A normal hospital glucose solution for patient injection is 5.00% glucose, $C_6H_{12}O_6$.

   (a) What is the mass of $C_6H_{12}O_6$ dissolved in 100.0 g of 5.00% glucose solution?

                                                    _____

   (b) What is the molar concentration of normal 5.00% glucose (180.18 g/mol) solution? (Assume the density of normal glucose solution is 1.02 g/mL.)

                                                    _____

Analysis of Vinegar

OBJECTIVES

- To standardize a sodium hydroxide solution with potassium hydrogen phthalate.
- To determine the molar concentration and mass/mass percent concentration of acetic acid in an unknown vinegar solution.
- To gain proficiency in the laboratory technique of titration.

DISCUSSION

In this experiment, we will neutralize an acidic solution of vinegar using a basic solution of sodium hydroxide. We will determine the amount of sodium hydroxide necessary by performing a **titration** using a buret. When the acid is completely neutralized by the base, the titration stops. This is called the **endpoint** and is signaled when an **indicator** changes color. The indicator in this experiment is phenolphthalein, which is colorless in acid and pink in base. At the endpoint in the titration, a single drop of base is sufficient to bring about a color change from colorless to pink. Figure 20.1 illustrates a typical titration.

We will begin the experiment by diluting $6 \, M$ NaOH with water. Since the dilution of NaOH provides only an approximate concentration, it is necessary to determine the concentration precisely by **standardization**. To standardize NaOH, we will weigh crystals of potassium hydrogen phthalate, $KHC_8H_4O_4$ (abbreviated KHP). After dissolving the KHP crystals in water, we will titrate the acid solution with NaOH according to the following equation.

$$KHP(aq) \; + \; NaOH(aq) \; \rightarrow \; KNaP(aq) \; + \; H_2O(l)$$

231

A 0.905 g sample of KHP (204.23 g/mol) is dissolved in water and titrated with 19.90 mL of NaOH solution to a phenolphthalein endpoint. Find the molarity of the NaOH solution.

Solution: Referring to the preceding equation for the reaction and applying the rules of stoichiometry, we have

$$0.905 \; \text{g KHP} \; \times \; \frac{1 \; \text{mol KHP}}{204.23 \; \text{g KHP}} \; \times \; \frac{1 \; \text{mol NaOH}}{1 \; \text{mol KHP}} \; = \; 0.00443 \; \text{mol NaOH}$$

The molarity of the NaOH is found as follows:

$$\frac{0.00443 \; \text{mol NaOH}}{19.90 \; \text{mL solution}} \; \times \; \frac{1000 \; \text{mL}}{1 \; \text{L}} \; = \; \frac{0.223 \; \text{mol NaOH}}{1 \; \text{L solution}} \; = \; 0.223 \; M \; \text{NaOH}$$

In this example, the concentration of the standard NaOH solution is 0.223 *M*.

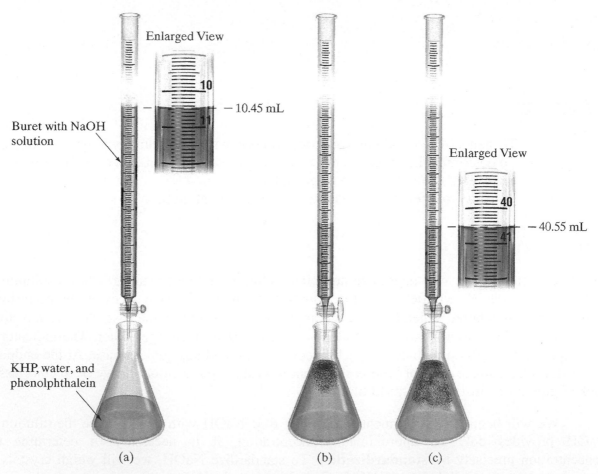

Figure 20.1 Apparatus for the Titration of an Acid with a Base
(a) Read the initial volume of NaOH in the buret (10.45 mL). (b) A flash of pink indicates an approaching endpoint. (c) A permanent pink color signals the final endpoint. Read the final volume of NaOH in the buret (40.55 mL). The volume of NaOH used for the titration is 40.55 mL − 10.45 mL = 30.10 mL.

After standardizing the sodium hydroxide solution, we can determine the concentration of acetic acid in an unknown vinegar solution. A sample of vinegar will be titrated with NaOH to a phenolphthalein endpoint. The equation for the reaction is

$$HC_2H_3O_2(aq) \ + \ NaOH(aq) \ \rightarrow \ NaC_2H_3O_2(aq) \ + \ H_2O(l)$$

The following example exercise illustrates the calculation for the percentage of acetic acid in an unknown vinegar sample.

Example Exercise 20.2 • Percentage of Acetic Acid in Vinegar

The titration of a 10.0-mL vinegar sample requires 29.05 mL of standard 0.223 M NaOH. Calculate the (a) molarity and (b) mass/mass percent concentration of acetic acid.

Solution: We can calculate the moles of acetic acid from the moles of NaOH solution:

$$29.05 \ \text{mL solution} \ \times \ \frac{0.223 \ \text{mol NaOH}}{1000 \ \text{mL solution}} \ \times \ \frac{1 \ \text{mol } HC_2H_3O_2}{1 \ \text{mol NaOH}} \ = \ 0.00648 \ \text{mol } HC_2H_3O_2$$

(a) The molar concentration of $HC_2H_3O_2$ is

$$\frac{0.00648 \ \text{mol } HC_2H_3O_2}{10.0 \ \text{mL solution}} \ \times \ \frac{1000 \ \text{mL}}{1 \ \text{L}} \ = \ \frac{0.648 \ \text{mol } HC_2H_3O_2}{1 \ \text{L solution}}$$

$$= \ 0.648 \ M \ HC_2H_3O_2$$

(b) To calculate the m/m % concentration, we must know the density of the vinegar (1.01 g/mL) and the molar mass of acetic acid (60.06 g/mol).

$$\frac{0.648 \ \text{mol } HC_2H_3O_2}{1000 \ \text{mL solution}} \ \times \ \frac{60.06 \ \text{g } HC_2H_3O_2}{1 \ \text{mol } HC_2H_3O_2} \ \times \ \frac{1 \ \text{mL solution}}{1.01 \ \text{g solution}} \ \times \ 100\%$$

$$= \ 3.85\% \ HC_2H_3O_2$$

- graduated cylinder
- 1000-mL Florence flask w/stopper
- 125-mL Erlenmeyer flasks (3)
- buret stand (or ring stand)
- buret clamp (or utility clamp)
- small, plastic funnel (optional)
- 50-mL buret
- 10-mL pipet and pipet bulb
- 100-mL beaker
- 150-mL beaker
- wash bottle with distilled water

- dilute sodium hydroxide,
  6 M NaOH
- potassium hydrogen phthalate,
  solid $KHC_8H_4O_4$ (KHP)
- phenolphthalein indicator
- unknown vinegar solution,
  3.00–5.00% $HC_2H_3O_2$

PROCEDURE

A. Preparation of Standard Sodium Hydroxide Solution

1. Half fill a 1000-mL Florence flask with ~500 mL of distilled water. Measure ~20 mL of 6 M NaOH into a graduated cylinder and pour the NaOH into the Florence flask. Stopper the flask, and carefully swirl to mix the solution.

2. Place the buret in a buret stand. Using a small funnel, half fill the buret with the NaOH solution from the Florence flask. Allow some solution to pass through the buret tip, invert the buret, and empty the remainder into the sink.

3. Close the stopcock, and fill the buret with NaOH solution from the Florence flask.

   Note: Carefully add NaOH solution to the funnel so as to not overfill the buret.

4. Label the 125-mL Erlenmeyer flasks #1, #2, and #3. Accurately weigh out ~1 g of KHP into each of the flasks. Add ~25 mL of distilled water to each flask, and heat as necessary to dissolve the KHP crystals.

5. Perform the titration as follows:
   - Drain some NaOH through the tip of the buret to clear any air bubbles.
   - Position Erlenmeyer flask #1 under the buret as shown in Figure 20.1.
   - Record the initial buret reading (± 0.05 mL).
   - Add a drop of phenolphthalein indicator to the flask.
   - Titrate the KHP sample to a permanent pink endpoint.
   - Record the final buret reading (± 0.05 mL).

6. Refill the buret with NaOH solution, record the initial buret reading, add a drop of phenolphthalein to flask #2, titrate the KHP sample, and record the final buret reading.

7. Refill the buret with NaOH solution, record the initial buret reading, add a drop of phenolphthalein to flask #3, titrate the KHP sample, and record the final buret reading.

8. Calculate the molarity of the NaOH solution for each trial. Record the average molarity of NaOH in the Data Table of Procedure B.

   Note: *Save the NaOH in the Florence flask for Procedure B.*

B. Concentration of Acetic Acid in Vinegar

1. Obtain ~50 mL of vinegar solution in a dry 100-mL beaker. Record the unknown number in the Data Table.

2. Condition a pipet with unknown vinegar solution, and transfer a 10.0-mL sample into each 125-mL flask (see **Appendix E**). Add ~25 mL of distilled water into each flask.

   Note: It is not necessary to use dry flasks.

3. Fill the buret with NaOH solution, adjust the meniscus to zero, and record the initial buret reading as 0.00 mL. Add a drop of phenolphthalein to flask #1 and titrate the vinegar sample to a pink endpoint. Record the final buret reading.

4. Refill the buret with NaOH solution, and adjust the meniscus to 0.00 mL. Add a drop of phenolphthalein to flask #2, and titrate the vinegar sample.

   Note: By adjusting the meniscus to 0.00 mL, the endpoints for samples #2 and #3 should be at the same final buret reading as sample #1.

5. Refill the buret with NaOH solution and adjust the meniscus to 0.00 mL. Add a drop of phenolphthalein to flask #3, and titrate the vinegar sample.

6. Calculate the molarity of acetic acid, $HC_2H_3O_2$, in the unknown vinegar solution.

7. Convert the molarity of $HC_2H_3O_2$ (60.06 g/mol) to mass/mass percent concentration. Assume the density is 1.01 g/mL for the unknown vinegar solution.

   Note: *When the titrations are complete, rinse the buret and all glassware with distilled water to remove traces of NaOH solution.*

DATE _____

NAME _____

SECTION _____

PRELABORATORY ASSIGNMENT*

1. In your own words, define the following terms:

   conditioning

   endpoint

   indicator

   molar concentration (M)

   mass/mass percent concentration (m/m %)

   standardization

   titration

2. Observe and record the following buret readings.

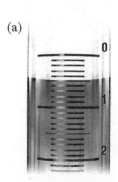
(a)

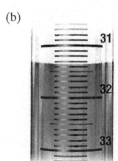
(b)

3. When approaching the endpoint for the titration, how can you tell that you are very close?

   What volume of NaOH is required to flip the indicator from colorless to pink at the endpoint?

4. If KHP sample #1 requires 27.30 mL of NaOH solution to reach an endpoint, what volume should be required for samples #2 and #3?

5. If vinegar sample #1 requires 30.15 mL of NaOH solution to reach an endpoint, what volume should be required for samples #2 and #3?

---

* Answers in Appendix J

6. A 1.015-g sample of KHP (204.23 g/mol) is dissolved in water and titrated with 24.55 mL of NaOH solution to a phenolphthalein endpoint. Refer to Example Exercise 20.1 and calculate the molarity of the NaOH solution.

7. A 10.0-mL vinegar sample is pipetted into an Erlenmeyer flask and titrated with 41.60 mL of 0.202 M NaOH to a phenolphthalein endpoint.

   (a) Refer to Example Exercise 20.2 and calculate the molarity of the acetic acid in the vinegar.

   (b) Assume the density of the vinegar solution is 1.01 g/mL and find the mass/mass percent concentration of acetic acid, $HC_2H_3O_2$ (60.06 g/mol), in the unknown vinegar sample.

8. Which of the following is a serious source of experimental error?

   (a) The sodium hydroxide is not mixed completely in the Florence flask.

   (b) The buret is not conditioned.

   (c) The KHP samples are dissolved in 50 mL (*not 25 mL*) of distilled water.

   (d) Two drops (*not one drop*) of phenolphthalein indicator is used.

   (e) Bubbles are not cleared from the tip of the buret.

   (f) The Erlenmeyer flasks are not dry before pipetting the vinegar samples.

9. What safety precautions should be observed in this experiment?

NAME _____

SECTION _____

DATA TABLE

A. Preparation of Standard Sodium Hydroxide Solution

mass of Erlenmeyer flask + KHP   _____ g   _____ g   _____ g

mass of Erlenmeyer flask   _____ g   _____ g   _____ g

mass of KHP   _____ g   _____ g   _____ g

final buret reading   _____ mL _____ mL _____ mL

initial buret reading   _____ mL _____ mL _____ mL

volume of NaOH   _____ mL _____ mL _____ mL

Show the calculation of the molarity of NaOH for trial 1 (see Example Exercise 20.1).

Molarity of NaOH   _____ M _____ M _____ M

Average molarity of NaOH   _____ M

B. Concentration of Acetic Acid in Vinegar   **UNKNOWN #** _____

     Average molarity of NaOH (see Procedure A)   _____ *M*

     volume of vinegar solution   _____ mL _____ mL _____ mL

     final buret reading   _____ mL _____ mL _____ mL

     initial buret reading   _____ mL _____ mL _____ mL

     volume of NaOH   _____ mL _____ mL _____ mL

Show the calculation for the molarity of acetic acid for trial 1 (see Example Exercise 20.2).

     Molarity of $HC_2H_3O_2$ _____ *M* _____ *M* _____ *M*

Show the calculation for the percent concentration of acetic acid for trial 1. (Assume the density of the vinegar solution is 1.01 g/mL.)

     Mass/mass percent $HC_2H_3O_2$ _____ % _____ % _____ %

        Average mass/mass percent $HC_2H_3O_2$ _____ %

NAME _____

SECTION _____

POSTLABORATORY ASSIGNMENT

1. A hydrochloric acid solution is standardized using 0.500 g of sodium carbonate, Na_2CO_3. Find the molarity of the acid if 29.50 mL are required to reach a phenolphthalein endpoint.

$$2\ HCl\,(aq)\quad +\quad Na_2CO_3\,(s)\quad \rightarrow\quad 2\ NaCl(aq)\quad +\quad H_2O(l)\quad +\quad CO_2\,(g)$$

_____

2. A 10.0-mL sample of household ammonia solution required 18.15 mL of 0.320 M HCl to reach neutralization. Calculate (a) the molar concentration of the ammonia and (b) the mass/mass percent concentration of ammonia (17.04 g/mol), given a solution density of 0.990 g/mL.

$$HCl\,(aq)\quad +\quad NH_3\,(aq)\quad \rightarrow\quad NH_4Cl\,(aq)$$

(a)_____

(b)_____

3. If 22.15 mL of 0.100 M sulfuric acid is required to neutralize 10.0 mL of lithium hydroxide solution, what is the molar concentration of the base?

$$H_2SO_4\,(aq)\quad +\quad 2\ LiOH(aq)\quad \rightarrow\quad Li_2SO_4\,(aq)\quad +\quad 2\ H_2O(l)$$

_____

4. A Rolaids tablet contains calcium carbonate, which neutralizes stomach acid. If a Rolaids tablet neutralizes 42.15 mL of 0.320 M hydrochloric acid, how many milligrams of calcium carbonate are in a Rolaids tablet?

$$CaCO_3(s) + 2 HCl(aq) \rightarrow CaCl_2(aq) + H_2O(l) + CO_2(g)$$

_____

5. (optional) A student carefully diluted 50.0 mL of 6 M HCl solution in 950.0 mL of distilled water. Calculate the molarity of the diluted acid solution.

_____

Explain why this diluted HCl solution *cannot* be used as a standard solution of acid.

Electrical Conductivity of Aqueous Solutions

- To observe the electrical conductivity of substances in aqueous solution.
- To determine whether an aqueous solution is a strong or weak electrolyte.
- To interpret a chemical reaction by observing aqueous solution conductivity.
- To become proficient in writing net ionic equations.

DISCUSSION

Electrical conductivity is based on the flow of electrons. Metals are good conductors of electricity because they allow electrons to flow through the metal. Distilled water is a very weak conductor because very little electricity passes through pure water. However, when a substance dissolves in water and forms ions, the ions are capable of conducting an electric current. If the substance is highly ionized, the solution is a strong conductor of electricity. If the substance is only slightly ionized, the solution is a weak conductor.

A substance with covalent bonds can dissolve in water to form positive and negative ions. For example, hydrogen chloride dissolves in water to form H^+ and Cl^- ions. The formation of positive and negative ions from a molecular compound, such as HCl, is called **ionization**.

A substance with ionic bonds can also dissolve in water to form positive and negative ions. For example, sodium chloride dissolves in water to form Na^+ and Cl^- ions. The separation of ions in an ionic compound is referred to as **dissociation**.

In this experiment, you will be testing conductivity using an apparatus that has two wires serving as electrodes (Figure 21.1). If the electrodes are immersed in a strong electrolyte solution, the circuit is completed and the light bulb in the apparatus glows brightly. If the electrodes are immersed in a weak electrolyte solution, the light bulb glows dimly.

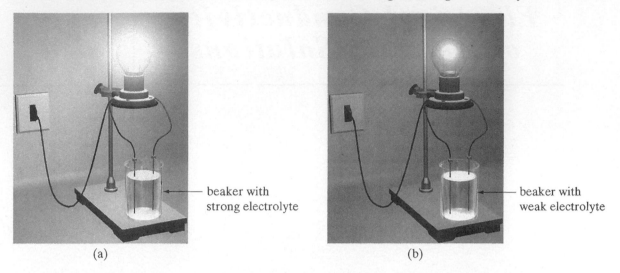

(a)   (b)

Figure 21.1 Apparatus for Conductivity Testing (a) A strong electrolyte is a
good conductor of electricity and the light bulb glows brightly. (b) A weak
electrolyte is a poor conductor of electricity and the light bulb glows dimly.

A solution that is a good conductor of an electric current is called a **strong electrolyte**. Examples of strong electrolytes include strong acids, strong bases, and salts that are highly soluble in aqueous solution. A solution that is a poor conductor of electricity is called a **weak electrolyte**. Examples of weak electrolytes include weak acids, weak bases, and salts that are only slightly soluble in aqueous solution. Table 21.1 lists several common examples of strong and weak electrolytes.

Table 21.1 Strong and Weak Electrolytes

Strong Electrolytes	Weak Electrolytes
Strong Acids	*Weak Acids*
hydrochloric acid, $HCl(aq)$	acetic acid, $HC_2H_3O_2(aq)$
nitric acid, $HNO_3(aq)$	carbonic acid, $H_2CO_3(aq)$
sulfuric acid, $H_2SO_4(aq)$	most other acids
Strong Bases	*Weak Bases*
sodium hydroxide, $NaOH(aq)$	ammonium hydroxide, $NH_4OH(aq)$
potassium hydroxide, $KOH(aq)$	most other bases
calcium hydroxide, $Ca(OH)_2(aq)$	
barium hydroxide, $Ba(OH)_2(aq)$	
Soluble Salts	*Very Slightly Soluble Salts*
sodium chloride, $NaCl(aq)$	silver chloride, $AgCl(s)$
potassium carbonate, $K_2CO_3(aq)$	calcium carbonate, $CaCO_3(s)$
cupric sulfate, $CuSO_4(aq)$	barium sulfate, $BaSO_4(s)$

Since strong electrolytes are highly ionized, we will indicate these substances as *ionized* in aqueous solution. Conversely, since weak electrolytes are only slightly ionized, we will indicate these substances as *nonionized* in aqueous solution. The following examples illustrate writing strong and weak electrolytes in aqueous solution.

Example 21.1 • Ionization of a Strong Acid

Sulfuric acid is a strong acid and the light bulb gives a bright glow when tested by the conductivity apparatus. Write H_2SO_4 as it exists in aqueous solution.

Solution: Sulfuric acid is a strong electrolyte that is highly ionized. Thus, we will write aqueous H_2SO_4 as *ionized*: $2 H^+(aq) + SO_4^{2-}(aq)$.

Example 21.2 • Ionization of a Weak Acid

Carbonic acid is a weak acid and the light bulb gives a dim glow when tested by the conductivity apparatus. Write H_2CO_3 as it exists in aqueous solution.

Solution: Carbonic acid is a weak electrolyte that is only slightly ionized. Thus, we will write aqueous H_2CO_3 as *nonionized*: $H_2CO_3(aq)$.

Example 21.3 • Ionization of a Strong Base

Potassium hydroxide is a strong base and the light bulb gives a bright glow when tested by the conductivity apparatus. Write KOH as it exists in aqueous solution.

Solution: Potassium hydroxide is a strong electrolyte that is highly ionized. Thus, we will write aqueous KOH as *ionized*: $K^+(aq) + OH^-(aq)$.

Example 21.4 • Ionization of a Weak Base

Ammonium hydroxide is a weak base and the light bulb gives a dim glow when tested by the conductivity apparatus. Write NH_4OH as it exists in aqueous solution.

Solution: Ammonium hydroxide is a weak electrolyte that is only slightly ionized. Thus, we will write aqueous NH_4OH as *nonionized*: $NH_4OH(aq)$.

Example 21.5 • Dissociation of a Soluble Salt

Aluminum chloride is a soluble salt and the light bulb gives a bright glow when tested by the conductivity apparatus. Write $AlCl_3$ as it exists in aqueous solution.

Solution: Aluminum chloride is a strong electrolyte that is highly ionized. Thus, we will write aqueous $AlCl_3$ as *ionized*: $Al^{3+}(aq) + 3 Cl^-(aq)$.

Writing Net Ionic Equations

Given the chemical equation for a reaction, balance the equation by inspection. Next, convert the balanced chemical equation into a **net ionic equation**, using the following guidelines.

1. Write a substance in the chemical equation in the ionized form if it is a strong electrolyte. Examples of strong electrolytes include: strong acids, strong bases, and soluble salts. Refer to Table 21.1 for the most common examples.

2. Write a substance in the chemical equation in the nonionized form if it is a weak electrolyte. Examples of weak electrolytes include weak acids, weak bases, insoluble salts, and water.

3. Write the **total ionic equation** that shows highly ionized substances in the ionic form and weakly ionized substances in the nonionized form.

4. Convert the total ionic equation to a net ionic equation by canceling **spectator ions**. Spectator ions must be identical on both sides of the total ionic equation.

5. Check the net ionic equation for (a) mass balance, and (b) ionic charge balance.

Example 21.6 • Double Replacement Net Ionic Equation

$$CaCl_2(aq) + K_2CO_3(aq) \rightarrow CaCO_3(s) + 2\,KCl(aq)$$

$$Ca^{2+}(aq) + 2\,\cancel{Cl^-}(aq) + 2\,\cancel{K^+}(aq) + CO_3^{2-}(aq) \rightarrow CaCO_3(s) + 2\,\cancel{K^+}(aq) + 2\,\cancel{Cl^-}(aq)$$

$$Ca^{2+}(aq) + CO_3^{2-}(aq) \rightarrow CaCO_3(s)$$

Example 21.7 • Double Replacement Net Ionic Equation

$$2\,AlBr_3(aq) + 3\,MgCl_2(aq) \rightarrow 2\,AlCl_3(aq) + 3\,MgBr_2(aq)$$

$$2\,\cancel{Al^{3+}}(aq) + 6\,\cancel{Br^-}(aq) + 3\,\cancel{Mg^{2+}}(aq) + 6\,\cancel{Cl^-}(aq) \rightarrow$$

$$2\,\cancel{Al^{3+}}(aq) + 6\,\cancel{Cl^-}(aq) + 3\,\cancel{Mg^{2+}}(aq) + 6\,\cancel{Br^-}(aq)$$

All spectator ions; thus, all the ions cancel and there is *No Reaction*.

Example 21.8 • Neutralization Net Ionic Equation

$$H_2SO_4(aq) + 2\,NaOH(aq) \rightarrow Na_2SO_4(aq) + 2\,H_2O(l)$$

$$2\,H^+(aq) + \cancel{SO_4^{2-}}(aq) + 2\,\cancel{Na^+}(aq) + 2\,OH^-(aq) \rightarrow 2\,\cancel{Na^+}(aq) + \cancel{SO_4^{2-}}(aq) + 2\,H_2O(l)$$

$$H^+(aq) + OH^-(aq) \rightarrow H_2O(l)$$

- conductivity apparatus
- small, dry beakers (6)
- straw (or 20 cm length of fire-polished glass tubing)
- wash bottle with distilled water
- concentrated acetic acid, $HC_2H_3O_2$
- sodium chloride, solid NaCl
- calcium carbonate, solid $CaCO_3$
- hydrochloric acid, 0.1 M HCl
- carbonic acid, 0.1 M H_2CO_3
- nitric acid, 0.1 M HNO_3
- sodium hydroxide, 0.1 M NaOH

- calcium hydroxide, sat'd $Ca(OH)_2$
- potassium iodide, 0.1 M KI
- aluminum nitrate, 0.1 M $Al(NO_3)_3$
- magnesium hydroxide, sat'd. $Mg(OH)_2$
- copper (II) sulfate, 0.1 M $CuSO_4$
- calcium nitrate, 0.1 M $Ca(NO_3)_2$
- potassium nitrate, 0.1 M KNO_3
- sodium carbonate, 0.1 M Na_2CO_3
- acetic acid, 0.1 M $HC_2H_3O_2$
- ammonium hydroxide, 0.1 M NH_4OH
- sulfuric acid, 0.1 M H_2SO_4
- barium hydroxide, 0.1 M $Ba(OH)_2$

PROCEDURE

A. Conductivity Testing—Evidence for Ions in Aqueous Solution

1. Test the conductivity of distilled water. Next, test the conductivity of tap water.

   Note: After each conductivity test, rinse the electrodes with distilled water. Record your observations in the Data Table, and state whether the conductivity test indicates a strong electrolyte or weak electrolyte. Write strong electrolytes as ions and weak electrolytes as molecules.

2. Pour about 10 mL of concentrated glacial acetic acid into a small dry beaker and test the conductivity. Add several milliliters of distilled water slowly to the acid while continuously testing the conductivity.

3. Place about 0.5 g of solid sodium chloride into a small dry beaker, and test the conductivity. Add distilled water to the NaCl and retest the conductivity of the solution.

4. Place about 0.5 g of solid calcium carbonate into a small dry beaker and test the conductivity. Add distilled water to the $CaCO_3$, and retest the conductivity of the solution.

5. Test the conductivity of each of the following solutions:
   (a) hydrochloric acid, 0.1 M HCl
   (b) carbonic acid, 0.1 M H_2CO_3
   (c) nitric acid, 0.1 M HNO_3
   (d) sodium hydroxide, 0.1 M NaOH
   (e) calcium hydroxide, saturated $Ca(OH)_2$
   (f) potassium iodide, 0.1 M KI
   (g) aluminum nitrate, 0.1 M $Al(NO_3)_3$
   (h) magnesium hydroxide, saturated $Mg(OH)_2$
   (i) copper (II) sulfate, 0.1 M $CuSO_4$
   (j) calcium nitrate, 0.1 M $Ca(NO_3)_2$

B. Conductivity Testing—Evidence for a Chemical Reaction

1. Test the conductivity of 0.1 M KNO_3 and 0.1 M Na_2CO_3 separately. Pour the solutions together, and retest the conductivity. Record your observations and conclusions. Balance the equation for the chemical reaction, and write the total ionic and net ionic equations.

2. Test the conductivity of 0.1 M $HC_2H_3O_2$ and 0.1 M NH_4OH separately. Pour the solutions together and retest the conductivity. Record your observations and conclusions. Balance the equation for the chemical reaction, and write the total ionic and net ionic equations.

3. Add 10 drops of 0.1 M H_2SO_4 into a beaker containing ~25 mL of water. Continuously test the conductivity of the solution while slowly adding 20 drops of $Ba(OH)_2$ dropwise. Record your observations and conclusions. Balance the equation for the chemical reaction, and write the total ionic and net ionic equations.

4. Add 10 drops of 0.1 M $Ba(OH)_2$ into a beaker containing ~25 mL of water. Continuously test the conductivity of the solution while blowing through a straw into the solution. Stop exhaling carbon dioxide, CO_2, when the conductivity is minimal. Record your observations and conclusions. Balance the equation for the chemical reaction, and write the total ionic and net ionic equations.

C. Net Ionic Equations—A Study Assignment

Balance the following chemical equations, and write the total ionic and net ionic equations. Refer to the section, *Writing Net Ionic Equations*, for directions and examples.

1. $FeCl_3(aq)$ + $Mg(NO_3)_2(aq)$ → $MgCl_2(aq)$ + $Fe(NO_3)_3(aq)$

2. $Ba(C_2H_3O_2)_2(aq)$ + $Na_2SO_4(aq)$ → $BaSO_4(s)$ + $NaC_2H_3O_2(aq)$

3. $H_2SO_4(aq)$ + $NH_4OH(aq)$ → $(NH_4)_2SO_4(aq)$ + $H_2O(l)$

4. $HNO_3(aq)$ + $Sr(OH)_2(aq)$ → $Sr(NO_3)_2(aq)$ + $H_2O(l)$

NAME _____

SECTION _____

PRELABORATORY ASSIGNMENT*

1. In your own words, define the following terms:

    dissociation

    ionization

    net ionic equation

    spectator ions

    strong electrolyte

    total ionic equation

    weak electrolyte

2. Explain the meaning of the following symbols:

    (g)

    (l)

    (s)

    (aq)

3. Give three examples for each of the following:

    (a) strong electrolyte

    (b) weak electrolyte

4. What will be observed when conductivity testing the following:

    (a) strong electrolyte

    (b) weak electrolyte

_____

Answers in Appendix J

5. Classify each of the following as ionized or nonionized; and write each substance as it exists in aqueous solution.

    (a) HBr(aq) — a strong acid

    (b) HF(aq) — a weak acid

    (c) $Sr(OH)_2$(aq) — a strong base

    (d) NH_4OH(aq) — a weak base

    (e) $AgNO_3$(aq) — a soluble salt

    (f) Ag_2SO_4(s) — an insoluble salt

6. Why must the electrodes on the conductivity apparatus, as well as all beakers, be rinsed with distilled water after each conductivity test?

7. What safety precautions should be observed in this experiment?

NAME _____

SECTION _____

DATA TABLE

A. Conductivity Testing—Evidence for Ions in Aqueous Solution

Solution	Observation	Conclusion	Ionized/Nonionized
LiOH (aq) HNO$_2$(aq)	bulb glows brightly bulb glows dimly	strong electrolyte weak electrolyte	Li$^+$(aq) + OH$^-$(aq) HNO$_2$(aq)
1. H$_2$O(l) – distilled H$_2$O – tap			— omit —
2. HC$_2$H$_3$O$_2$(l) HC$_2$H$_3$O$_2$(aq)			
3. NaCl (s) NaCl (aq)			
4. CaCO$_3$(s) CaCO$_3$(aq)			

Solution	Observation	Conclusion	Ionized/Nonionized
5. (a) HCl (aq)			
(b) H_2CO_3(aq)			
(c) HNO_3(aq)			
(d) NaOH (aq)			
(e) $Ca(OH)_2$(aq)			
(f) KI (aq)			
(g) $Al(NO_3)_3$(aq)			
(h) $Mg(OH)_2$(aq)			
(i) $CuSO_4$(aq)			
(j) $Ca(NO_3)_2$(aq)			

B. Conductivity Testing—Evidence for a Chemical Reaction

Solution	Observation	Conclusion

1. (a) $KNO_3(aq)$ _____ _____

 (b) $Na_2CO_3(aq)$ _____ _____

 (c) $KNO_3(aq) + Na_2CO_3(aq)$ _____ _____

equation: $KNO_3(aq)$ + $Na_2CO_3(aq)$ → $K_2CO_3(aq)$ + $NaNO_3(aq)$

total ionic:

net ionic:

2. (a) $HC_2H_3O_2(aq)$ _____ _____

 (b) $NH_4OH(aq)$ _____ _____

 (c) $HC_2H_3O_2(aq) + NH_4OH(aq)$ _____ _____

equation: $HC_2H_3O_2(aq)$ + $NH_4OH(aq)$ → $NH_4C_2H_3O_2(aq)$ + $H_2O(l)$

total ionic:

net ionic:

3. (a) $H_2SO_4(aq)$ _____ _____

 (b) $Ba(OH)_2(aq)$ _____ _____

 (c) $H_2SO_4(aq) + Ba(OH)_2(aq)$ _____ _____

equation: $H_2SO_4(aq)$ + $Ba(OH)_2(aq)$ → $BaSO_4(s)$ + $H_2O(l)$

total ionic:

net ionic:

4. (a) $Ba(OH)_2(aq)$ _____ _____

 (b) $CO_2(g)$ _____ _____

 (c) $Ba(OH)_2(aq) + CO_2(g)$ _____ _____

equation: $Ba(OH)_2(aq)$ + $CO_2(g)$ → $BaCO_3(s)$ + $H_2O(l)$

total ionic:

net ionic:

C. Net Ionic Equations—A Study Assignment

1. equation: $FeCl_3(aq)$ + $Mg(NO_3)_2(aq)$ → $MgCl_2(aq)$ + $Fe(NO_3)_3(aq)$

   total ionic:

   net ionic:

---

2. equation: $Ba(C_2H_3O_2)_2(aq)$ + $Na_2SO_4(aq)$ → $BaSO_4(s)$ + $NaC_2H_3O_2(aq)$

   total ionic:

   net ionic:

---

3. equation: $H_2SO_4(aq)$ + $NH_4OH(aq)$ → $(NH_4)_2SO_4(aq)$ + $H_2O(l)$

   total ionic:

   net ionic:

---

4. equation: $HNO_3(aq)$ + $Sr(OH)_2(aq)$ → $Sr(NO_3)_2(aq)$ + $H_2O(l)$

   total ionic:

   net ionic:

POSTLABORATORY ASSIGNMENT

1. Why is distilled water a weaker conductor of electricity than tap water? (Refer to **A.1** in the Data Table.)

_____

2. Why does concentrated glacial acetic acid first act as a nonconductor and then behave as a weak conductor with the addition of water? (Refer to **A.2** in the Data Table.)

_____

3. Why does solid sodium chloride act as a nonelectrolyte while an aqueous NaCl solution acts as a strong electrolyte? (Refer to **A.3** in the Data Table.)

_____

4. Why do aqueous solutions of $HC_2H_3O_2$ and NH_4OH act as weak electrolytes individually, but after they are mixed together, they act as a strong electrolyte? (Refer to **B.2** in the Data Table.)

_____

5. Why do aqueous solutions of H_2SO_4 and $Ba(OH)_2$ act as strong electrolytes individually, but after they are mixed together, they act as a weak electrolyte? (Refer to **B.3** in the Data Table.)

_____

6. Why does blowing carbon dioxide gas into aqueous barium hydroxide, $Ba(OH)_2$, reduce the conductivity from a strong electrolyte to a weak electrolyte? (Refer to **B.4** in the Data Table.)

_____

7. State whether each of the following aqueous solutions is highly ionized or slightly ionized.

   (a) $H_3PO_4(aq)$ (b) $Ca(OH)_2(aq)$

   (c) $CoCrO_4(aq)$ (d) $Li_2CO_3(aq)$

8. Balance the chemical equations below, and then write the total ionic and net ionic equations. Designate the state of each species by (s), (l), (g), or (aq).

   (a) $ZnCl_2(aq)$ + $Na_2CO_3(aq)$ → $ZnCO_3(s)$ + $NaCl\,(aq)$

         total ionic:

         net ionic:

   (b) $AlCl_3(aq)$ + $NH_4OH\,(aq)$ → $Al(OH)_3(s)$ + $NH_4Cl\,(aq)$

         total ionic:

         net ionic:

   (c) $HC_2H_3O_2(aq)$ + $Ca(OH)_2(aq)$ → $Ca(C_2H_3O_2)_2(aq)$ + $H_2O(l)$

         total ionic:

         net ionic:

9. (optional) Write a balanced chemical equation for the reaction of aqueous sodium nitrate and aqueous calcium acetate. Show the total ionic and net ionic equations.

Activity Series of Metals

OBJECTIVES

- To observe the oxidation numbers for an element in a compound or ion.
- To be able to write balanced chemical equations for redox reactions.
- To determine the placement of an unknown metal (**X**) in an activity series.

DISCUSSION

An **oxidation number** is the positive or negative number that describes the combining capacity of an element in a compound or polyatomic ion. Elements that are not combined with another element are assigned a value of zero. For example, zinc metal and oxygen gas each have a value of zero. However, in zinc oxide (ZnO), zinc has an oxidation number of positive two (Zn^{2+}), and oxygen negative two (O^{2-}). The following examples illustrate the calculation of oxidation numbers.

Example Exercise 22.1 • Oxidation Number of C in a Compound

Calculate the oxidation number of carbon in sodium hydrogen carbonate, $NaHCO_3$.

Solution: The oxidation number (ox no) of sodium ion is +1, of a hydrogen ion, +1, and that of oxygen, −2. Since the sum of all oxidation numbers for the elements in a compound must equal zero, we can write the following equation.

$$+1 +1 + \text{ox no C} + 3(-2) = 0$$
$$+2 + \text{ox no C} - 6 = 0$$
$$\text{ox no C} = +4$$

257

Example Exercise 22.2 • Oxidation Number of C in a Polyatomic Ion

Calculate the oxidation number of carbon in the oxalate ion, $C_2O_4^{2-}$.

Solution: The oxidation number of each oxygen atom is –2. The polyatomic ion has a charge of –2 and there are two carbon atoms; therefore,

$$2(\text{ox no C}) + 4\,(-2) = -2$$
$$2(\text{ox no C}) - 8 = -2$$
$$2(\text{ox no C}) = -2 + 8 = +6$$
$$\text{ox no C} = +3$$

In summary, we can calculate an oxidation number using the following guidelines.*

- A metal or a nonmetal in the free state has an oxidation number of 0.
- A monoatomic ion has an oxidation number equal to its ionic charge.
- A hydrogen atom is usually assigned an oxidation number of +1.
- An oxygen atom is usually assigned an oxidation number of –2.
- For a molecular compound, the more electronegative element is assigned a negative oxidation number equal to its charge as an anion.
- For an ionic compound, the sum of the oxidation numbers for each atom is equal to 0.
- For a polyatomic ion, the sum of the oxidation numbers for each atom is equal to the ionic charge on the polyatomic ion.

Balancing Redox Equations

A process in which a substance undergoes an increase in oxidation number by losing electrons is called **oxidation**. The opposite process in which a substance undergoes a decrease in oxidation number by gaining electrons is called **reduction**. To balance a redox reaction, we can use either of two methods: the oxidation number method or the ion-electron method. The ion-electron method is also called the half-reaction method. Exercise 22.3 illustrates the *oxidation number method*, and Example Exercise 22.4 illustrates the *ion-electron method*.

Example Exercise 22.3 • Balancing Equations by Oxidation Numbers

Balance the following equation by the oxidation number method. The reaction takes place in acidic solution.

$$Ag + NO_3^- \rightarrow Ag^+ + NO$$

Solution: First, note the change in oxidation number for silver: 0 to +1. Silver loses 1e–. Second, note the change in oxidation number for nitrogen: +5 to +2. Nitrogen gains 3e–. To balance electrons, the coefficient of silver is three; thus, 3 Ag and 3 Ag+. The coefficient of each nitrogen species is one; thus,

$$3\,Ag + NO_3^- \rightarrow 3\,Ag^+ + NO$$

The next step is to balance the oxygen atoms, using water molecules.

$$3\,Ag + NO_3^- \rightarrow Ag^+ + NO + 2\,H_2O$$

---

* There are a few exceptions to these general rules; however, none of the exceptions appears in this experiment.

Since the reaction takes place in acidic solution, we will balance the hydrogen atoms using hydrogen ions.

$$4\,H^+ + 3\,Ag + NO_3^- \rightarrow 3\,Ag^+ + NO + 2\,H_2O$$

The last step is to check the equation for balance. A redox equation must be balanced in terms of the number of atoms. Also check that the total charge of the reactants and products are equal.

$$4\,H^+ + 3\,Ag + NO_3^- \rightarrow 3\,Ag^+ + NO + 2\,H_2O$$

The check reveals that the atoms of reactants and products are balanced. Furthermore, the total ionic charge of the reactants is +3; the total product charge is also +3.

Example Exercise 22.4 • Balancing Redox Equations by Half-Reactions

Balance the redox reaction in the previous example exercise using the ion-electron method; that is, the half-reaction method.

$$Ag + NO_3^- \rightarrow Ag^+ + NO \quad (in\ acid)$$

Solution: Using this method, the oxidation reaction is treated separately from the reduction reaction although each can take place only in conjunction with the other. The two half-reactions are then added together to give the balanced redox equation. First, write the partial equation for each half-reaction.

$$Ag \rightarrow Ag^+$$
$$NO_3^- \rightarrow NO$$

Balance the partial equation, using water and hydrogen ions.

$$Ag \rightarrow Ag^+$$
$$4\,H^+ + NO_3^- \rightarrow NO + 2\,H_2O$$

Next, write the two half-reactions, and balance the charge using electrons.

oxidation: $\qquad\qquad Ag \rightarrow Ag^+ + e^-$
reduction: $\ 4\,H^+ + NO_3^- + 3\,e^- \rightarrow NO + 2\,H_2O$

The reduction half-reaction gains $3\,e^-$ for every $1\,e^-$ lost in the oxidation half-reaction. Therefore, we will multiply the oxidation half-reaction by three.

$$3\,Ag \rightarrow 3\,Ag^+ + 3\,e^-$$

Let's add the two half-reactions together to obtain the overall redox equation. After canceling $3\,e^-$ from each side of the equation, we have

$$4\,H^+ + 3\,Ag + NO_3^- \rightarrow 3\,Ag^+ + NO + 2\,H_2O$$

The number of atoms of reactants and products are equal. Also notice that the total ionic charge is +3 for both reactants and products.

An **activity series** is a list of metals arranged in order of their ability to displace another metal from aqueous solution. A metal that is more active will reduce another metal ion to a metal in the free state. For example, mercury is more active than gold and can reduce Au^{3+} to Au.

$$Hg(l) + Au^{3+}(aq) \rightarrow Hg^{2+}(aq) + Au(s)$$

On the other hand, the reverse process gives no reaction. That is, gold cannot reduce Hg^{2+} to Hg, because gold is less active than mercury, according to the activity series.

$$Au(s) + Hg^{2+}(aq) \rightarrow NR$$

For reference purposes, hydrogen (H) is included in the series. Metals above hydrogen in the activity series displace hydrogen gas from acid solutions. Metals below hydrogen do not react with dilute acids. Example Exercise 22.5 illustrates the determination of a partial activity series.

Example Exercise 22.5 • Determining an Activity Series

Determine the activity series for iron, cadmium, mercury, and hydrogen, based on the following observations.

1. An iron nail develops a shiny metallic surface when placed in a cadmium solution. The equation is

$$Fe(s) + Cd^{2+}(aq) \rightarrow Fe^{2+}(aq) + Cd(s)$$

2. The shiny metallic cadmium reacts with dilute acid according to the equation

$$Cd(s) + 2 H^{+}(aq) \rightarrow Cd^{2+}(aq) + H_2(g)$$

3. The dilute acid does not react with liquid mercury. Thus,

$$Hg(l) + H^{+}(aq) \rightarrow NR$$

Solution: Since iron displaces cadmium from aqueous solution, we know that Fe must be higher in the activity series (Fe > Cd).

Cadmium displaces hydrogen gas from acid solution; thus, Cd is higher than hydrogen in the series (Cd > H).

Since liquid mercury does not react with dilute acid, Hg is below hydrogen in the series (H > Hg).

In summary, we can conclude that the activity series is

Fe > Cd > (H) > Hg

most active   *least active*

- 13 × 100 mm test tubes (6) and test tube rack
- test tube brush
- wash bottle with distilled water
- evaporating dish
- iron wire, Fe
- ferric oxide, powder, Fe_2O_3
- ferric nitrate, 0.5 M $Fe(NO_3)_3$
- stannous chloride, 0.5 M $SnCl_2$
- manganese chips, Mn
- potassium permanganate, 0.010 M $KMnO_4$
- dilute sodium hydroxide, 6 M NaOH
- sodium sulfite, 0.5 M Na_2SO_3
- dilute hydrochloric acid, 6 M HCl

- sulfur, S powder
- sodium sulfide, solid Na_2S
- dilute sulfuric acid, 3 M H_2SO_4
- iodine solution, 0.5 M I_2/KI
- sodium thiosulfate, 0.1 M $Na_2S_2O_3$
- dilute nitric acid, 6 M HNO_3
- copper, Cu metal
- concentrated nitric acid, 16 M HNO_3
- ammonium chloride, solid NH_4Cl
- sodium sulfite, 0.5 M Na_2SO_3
- magnesium, Mg metal
- zinc, Zn metal
- zinc sulfate, 0.1 M $ZnSO_4$
- silver nitrate, 0.1 M $AgNO_3$
- unknown metal samples (**X**)

PROCEDURE

A. Oxidation Numbers of Iron

1. Inspect a piece of iron wire. Record your observations in the Data Table, and state the oxidation number for iron wire, Fe.

2. Inspect a sample of ferric oxide powder. Record your observations in the Data Table, and calculate the oxidation number for iron in Fe_2O_3.

3. Put 1 mL of aqueous ferric nitrate into a test tube. Note the color, and calculate the oxidation number for iron in $Fe(NO_3)_3$. Add 1 mL of aqueous stannous chloride to the test tube, and observe the color change as the ferrous ion is formed in solution. Record your observations, and state the oxidation number for iron in Fe^{2+}.

B. Oxidation Numbers of Manganese

1. Examine a piece of manganese metal. Record your observations in the Data Table, and state the oxidation number for manganese metal, Mn.

2. Place 2 mL of potassium permanganate in a clean test tube. Add a drop of dilute hydrochloric acid and then several drops of sodium sulfite to produce the manganese ion. Record your observations, and state the oxidation number for manganese in Mn^{2+}.

3. Introduce 2 mL of potassium permanganate solution into a clean test tube. Add sodium sulfite drop by drop until the purple color fades. After several minutes, observe the solid particles of manganese dioxide forming in the solution. Record your observations, and calculate the oxidation number for manganese in MnO_2.

4. Deliver 2 mL of potassium permanganate into a test tube. Note the color, and calculate the oxidation number for manganese in the permanganate ion, MnO_4^-. Add a few drops of dilute sodium hydroxide and one drop of sodium sulfite into the test tube. Observe the color change and the formation of the manganate ion. Calculate the oxidation number for manganese in the manganate ion, MnO_4^{2-}.

C. Oxidation Numbers of Sulfur

1. Examine a small portion of powdered sulfur. Record your observations in the Data Table, and state the oxidation number for sulfur, S.

2. Place a pea-sized portion of sulfur in an evaporating dish, and ignite it with a burner. Notice the color of the flame and the white sulfur dioxide gas. Calculate the oxidation number for sulfur in gaseous SO_2.

   Caution: Perform this operation under a fume hood. Avoid breathing the sulfur dioxide gas, and dispose of the test tube contents under the fume hood.

3. Put a very small crystal of sodium sulfide, Na_2S, in a dry test tube. Add several drops of dilute sulfuric acid, H_2SO_4. Describe the odor of the hydrogen sulfide, H_2S, gas given off. Calculate the oxidation number for sulfur in Na_2S, H_2SO_4, and H_2S.

   Caution: Perform this operation under a fume hood. Avoid breathing H_2S, but waft the gas briefly to note its odor. Dispose of the test tube contents under the fume hood.

4. Pour about 2 mL of iodine solution into a test tube. Add sodium thiosulfate, $Na_2S_2O_3$, until the iodine is discolored. Record the change, and calculate the oxidation number for sulfur in $Na_2S_2O_3$.

D. Oxidation Numbers of Nitrogen

1. Air contains over 78% nitrogen. By simply breathing the air, note the odor and taste, and observe the color. State the oxidation number for nitrogen, N_2, in air.

2. Put 2 mL of *dilute* nitric acid, HNO_3, in a test tube. Add a piece of copper metal, and observe the reaction that releases nitrogen monoxide, NO. Record your observations, and calculate the oxidation number for nitrogen in HNO_3 and NO.

3. Deliver 2 mL of *concentrated* nitric acid, HNO_3, into a test tube. Add a piece of copper metal, and observe the reaction that evolves nitrogen dioxide gas, NO_2. Record your observations, and calculate the oxidation number for nitrogen in NO_2.

   Caution: Perform this operation under a fume hood. Concentrated HNO_3 should be handled with great caution. Avoid breathing the fumes, and dispose of the test tube contents under the fume hood.

4. Using a spatula, introduce a pea-sized portion of solid ammonium chloride, NH_4Cl, into a clean test tube. Add a dropper of dilute sodium hydroxide, and cover the end of the test tube with your thumb for 30 seconds. Shake the test tube, release your thumb, and waft the gas to detect the odor. Calculate the oxidation number for nitrogen in ammonium chloride, NH_4Cl, and ammonia gas, NH_3.

E. Oxidation-Reduction Equations

Many of the reactions in Procedures A–D illustrate oxidation–reduction and are listed in the Data Table. Balance each redox equation using either the oxidation number method or the ion-electron (half-reaction) method.

F. Activity Series and an Unknown Metal

1. Obtain an unknown metal (**X**), and record the number in the Data Table.

2. Add 2 mL of dilute hydrochloric acid to each of four test tubes. Drop a small piece of Cu, Mg, Zn, or **X** into each test tube. Record your observations in the Data Table.

3. Clean the four test tubes, and then add 2 mL of zinc sulfate solution. Put a small piece of Cu, Mg, Zn, or **X** into each test tube. Record your observations in the Data Table.

4. Clean the four test tubes, and then add 2 mL of silver nitrate solution. Put a small piece of Cu, Mg, Zn, or **X** into each test tube. Record your observations.

5. Based on the foregoing observations, list the experimental activity series for the following: Cu, Mg, Zn, Ag, (H), and **X**.

EXPERIMENT 22 NAME ______________________

DATE ______________ SECTION ______________________

PRELABORATORY ASSIGNMENT*

1. In your own words, define the following terms:

   activity series

   oxidation

   oxidation number

   oxidizing agent

   redox reaction

   reducing agent

   reduction

2. Calculate the oxidation number of chlorine in each of the following.

   (a) Cl_2

   (b) HCl

   (c) $NaClO$

   (d) ClO_4^-

   (e) Cl_2O_5

---

* *Answers in Appendix J*

3. Balance the following oxidation–reduction equations.

   (a) $Cr_2O_7^{2-}$ + Fe^{2+} + H^+ → Cr^{3+} + Fe^{3+} + H_2O

   (b) Mn^{2+} + H_2O_2 + OH^- → MnO_2 + H_2O

4. Assume the following reactions go essentially to completion:

$$Al + Cd^{2+} \rightarrow Al^{3+} + Cd$$

$$Ni + Ag^+ \rightarrow Ni^{2+} + Ag$$

$$Cd + Ni^{2+} \rightarrow Cd^{2+} + Ni$$

   Write the activity series for the metals Al, Cd, Ni, and Ag. List the most active metal first.

5. What safety precautions must be observed in this experiment?

NAME _____

SECTION _____

DATA TABLE

A. Oxidation Numbers of Iron

Substance	Observation	Oxidation Number
1. Fe		_____
2. Fe_2O_3		_____
3. $Fe(NO_3)_3$		_____
Fe^{2+}		_____

B. Oxidation Numbers of Manganese

Substance		Oxidation Number
1. Mn		_____
2. Mn^{2+}		_____
3. MnO_2		_____
4. MnO_4^-		_____
MnO_4^{2-}		_____

C. Oxidation Numbers of Sulfur

   1. S   _____

   2. SO_2 _____

   3. Na_2S _____

      H_2SO_4 _____

      H_2S _____

   4. $Na_2S_2O_3$ _____

D. Oxidation Numbers of Nitrogen

   1. N_2 _____

   2. HNO_3 _____

      NO   _____

   3. NO_2 _____

   4. NH_4Cl _____

      NH_3 _____

E. Oxidation–Reduction Equations

1. Reduction of Fe^{3+} in neutral solution

$$Fe^{3+} \ + \ Sn^{2+} \ \rightarrow \ Fe^{2+} \ + \ Sn^{4+}$$

2. Reduction of MnO_4^- in acidic solution

$$MnO_4^- \ + \ SO_3^{2-} \ \rightarrow \ Mn^{2+} \ + \ SO_4^{2-}$$

3. Reduction of MnO_4^- in neutral solution

$$MnO_4^- \ + \ SO_3^{2-} \ \rightarrow \ MnO_2 \ + \ SO_4^{2-}$$

4. Reduction of MnO_4^- in basic solution

$$MnO_4^- \ + \ SO_3^{2-} \ \rightarrow \ MnO_4^{2-} \ + \ SO_4^{2-}$$

5. Oxidation of $S_2O_3^{2-}$ in neutral solution

$$S_2O_3^{2-} \ + \ I_2 \ \rightarrow \ S_4O_6^{2-} \ + \ I^-$$

6. Oxidation of Cu in dilute acid

$$Cu \ + \ NO_3^- \ \rightarrow \ Cu^{2+} \ + \ NO$$

7. Oxidation of Cu in concentrated acid

$$Cu \ + \ NO_3^- \ \rightarrow \ Cu^{2+} \ + \ NO_2$$

F. Activity Series and an Unknown Metal   **UNKNOWN #** _____

	Cu	Mg	Zn	X
HCl				
ZnSO$_4$				
AgNO$_3$				

1. Based on the reactions with hydrochloric acid, which metals are more active than hydrogen (H)?

2. Based on the reactions with zinc sulfate, which metals are more active than zinc?

3. Based on the reactions with silver nitrate, which metals are more active than silver?

Activity Series for Cu, Mg, Zn, Ag, (H), and X:

_____   _____   _____   _____   _____   _____

most   *least*
active   *active*

POSTLABORATORY ASSIGNMENT

1. Calculate the oxidation number for iodine in each of the following.

    (a) KIO_4

    (b) HI

    (c) IF_7

    (d) NaIO

    (e) IO_2^-

    (f) I_2O_5

    (g) IO_3^-

    _____

    (h) I_2

    _____

2. Balance the following oxidation–reduction equations. Assume each reaction takes place in acidic solution.

    (a) $MnO_4^- + Fe^{2+} \rightarrow Mn^{2+} + Fe^{3+}$

    (b) $Sn^{2+} + IO_3^- \rightarrow Sn^{4+} + I^-$

    (c) $NO_3^- + H_2S \rightarrow NO + S$

    (d) $Mn^{2+} + BiO_3^- \rightarrow MnO_4^- + Bi^{3+}$

    (e) $Cr_2O_7^{2-} + C_2O_4^{2-} \rightarrow Cr^{3+} + CO_2$

3. Balance the following oxidation–reduction equations. Assume each reaction takes place in basic solution.

(a) SO_3^{2-} + Cl_2 → SO_4^{2-} + Cl^-

(b) MnO_2 + O_2 → MnO_4^{2-} + H_2O

(c) MnO_4^- + BrO_2^- → MnO_2 + BrO_4^-

4. Consider the following redox equation:

$$I_2 + 5\,Cl_2 + 5\,H_2O → 2\,IO_3^- + 10\,Cl^- + 12\,H^+$$

Identify the species undergoing

(a) oxidation _____ (b) reduction _____

Identify the species functioning as the

(a) oxidizing agent _____ (b) reducing agent _____

5. Solutions of bromide, fluoride, and iodide were treated with bromine water. Iodide was oxidized to iodine; bromide and fluoride did not react. Arrange the three elements Br_2, F_2, and I_2 in order of their ability to oxidize another species.

Strongest oxidizing agent: _____ _____ _____

6. (optional) Certain substances demonstrate the ability to undergo oxidation and reduction simultaneously. This phenomenon is called *disproportionation*. For example, chlorine in a basic hydroxide solution produces chlorate and chloride ions. Write a balanced redox equation for the disproportionation reaction.

Organic Models and Functional Groups

- To build molecular models for the following hydrocarbons: alkanes, alkenes, alkynes, and arenes.
- To build molecular models for the following hydrocarbon derivatives: organic halides, alcohols, phenols, ethers, amines, aldehydes, ketones, carboxylic acids, esters, and amides.
- To identify the class of compound for unknown model structures.

Organic chemistry is the study of compounds that contain the element carbon. Inorganic chemistry is the study of compounds that do not contain carbon. Interestingly, the element carbon is found in over seven million different compounds. There are two reasons why over 90% of all compounds are organic. First, carbon is unusual in that it has the ability to self-link, forming chains of carbon atoms. Second, organic compounds typically contain several carbon atoms that may form isomers by joining in more than one arrangement or configuration.

Compounds having the same molecular formula but different structural formulas are called **isomers**. For instance, the molecular formula C_4H_{10} may be constructed in two ways and will satisfy the bond requirements for carbon (four bonds) and hydrogen (one bond). Figure 23.1 illustrates the isomers of butane, C_4H_{10}. Although the entire molecule as well as the individual bonds can be rotated in space to give what appears to be additional structures for the formula C_4H_{10}, careful examination will reveal that there are only two possibilities.

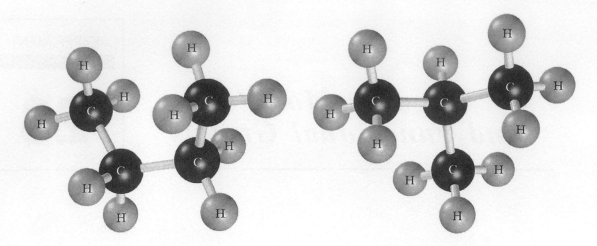

Figure 23.1 Structural Isomers The two isomers share the molecular formula C_4H_{10}; butane is shown on the left and "isobutane" on the right.

There are millions of organic compounds that may be cataloged into a few families. A **hydrocarbon** has only hydrogen and carbon atoms, and may be classified as an alkane, alkene, alkyne, or arene. A **hydrocarbon derivative** is related to a hydrocarbon and may contain oxygen, nitrogen, or a halogen. A **functional group** characterizes each family, and imparts similar properties to a **class of compounds**. Figure 23.2 illustrates an overall classification scheme for organic compounds.

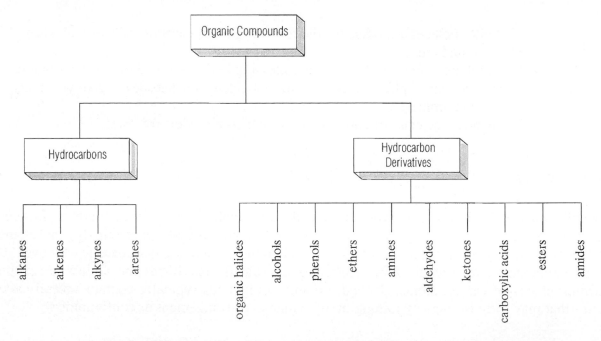

Figure 23.2 Classes of Organic Compounds Organic compounds can be organized systematically into four classes of hydrocarbons, and ten classes of hydrocarbon derivatives as shown above.

In this experiment, we will build example molecular models for each of the hydrocarbons and derivatives shown in Figure 23.2. The following examples will serve to correlate the classes of compounds with the molecular models you will construct.

Class of Compound	Example	Model Representation
alkane	CH_3—CH_3	
alkene	$CH_2 = CH_2$	
alkyne	$HC \equiv CH$	
arene	C_6H_6	

Substituting a chlorine, bromine, or iodine atom for a hydrogen atom onto the hydrocarbon chain produces a class of compounds called the *organic halides*.

organic halide   CH₃—Cl

Many organic compounds, especially those of biological interest, contain oxygen as well as carbon and hydrogen. Oxygen has a bond requirement of two, and this may be satisfied in a number of ways. If one bond is attached to a carbon and the second bond to a hydrogen, an *alcohol* is formed.

alcohol   CH₃CH₂—OH

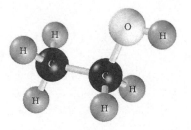

A *phenol* is simply a special example of an alcohol where the –OH is attached directly to a benzene ring.

phenol   C₆H₅—OH

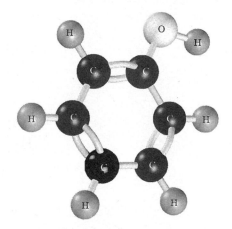

An oxygen atom may also be bonded to two carbon atoms. If that is the case, the class of compound is an *ether*.

ether   CH₃—O—CH₃

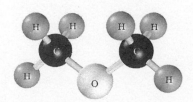

If we attach the –NH$_2$ group directly to a hydrocarbon, an *amine* results. Low-molecular-weight amines smell like ammonia and develop a fishy smell as the hydrocarbon chain increases.

amine   CH$_3$—NH$_2$

Oxygen can also satisfy its bond requirement of two by forming a double bond with carbon; this is called a **carbonyl group**.

$$\begin{matrix} O \\ \| \\ -C- \end{matrix}$$

If the carbonyl group is at the end of the molecule—that is, the carbonyl is bonded to a hydrogen—the functional group is an *aldehyde*. If the carbonyl group is in the middle of the carbon chain—that is, the carbonyl is bonded to two other carbon atoms—the functional group is a *ketone*.

aldehyde $$CH_3CH_2-\overset{\displaystyle O}{\overset{\|}{C}}-H$$

ketone $$CH_3-\overset{\displaystyle O}{\overset{\|}{C}}-CH_3$$

Another possibility is that the carbonyl group may be attached to a –OH group. The resulting structure is found in the class of compounds called *carboxylic acids*.

carboxylic acid

$$CH_3 - \overset{\overset{\displaystyle O}{\|}}{C} - OH$$

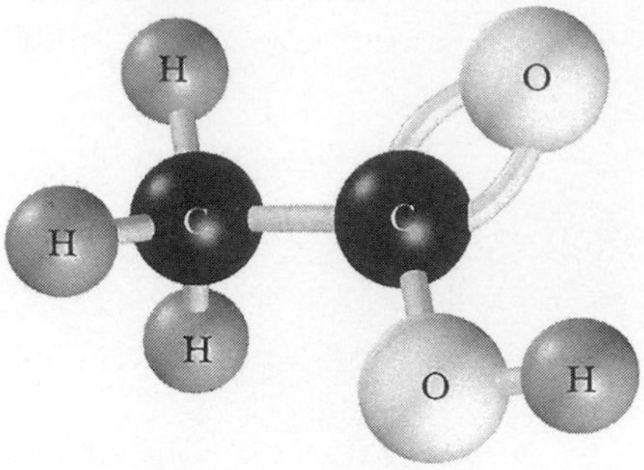

If we substitute a carbon atom for the hydrogen in the carboxylic acid group, we have a class of compounds called *esters*. Esters are noted for their typically fragrant odors.

ester

$$H - \overset{\overset{\displaystyle O}{\|}}{C} - O - CH_3$$

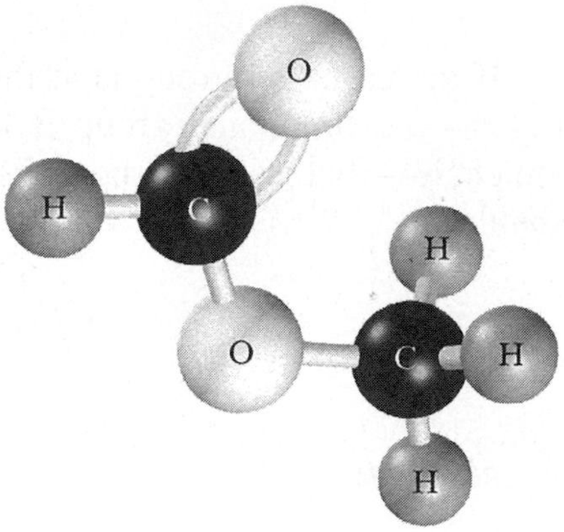

Starting with a carboxylic acid, we can remove the –OH group and replace it with –NH$_2$. The resulting class of compound is called an *amide*.

amide

$$H - \overset{\overset{\displaystyle O}{\|}}{C} - NH_2$$

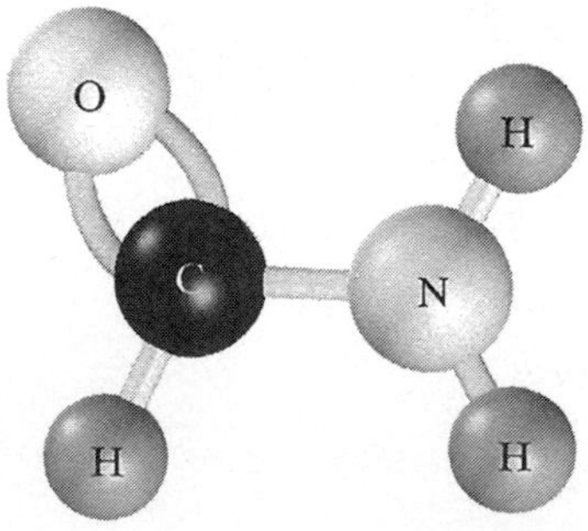

- Molecular Model Kit
  Student molecular model sets (ISBN: 0-205-08136-3) are available
  from Prentice Hall @ 1-800-922-0579 (www.prenhall.com).

Directions for Using Molecular Models

When constructing a model, a rigid connector between two balls represents a single bond. If two balls are joined by two flexible connectors, the two connectors represent a double bond. If two balls are joined by three connectors, the three connectors represent a triple bond.

one rigid connector	—	single bond
two flexible connectors	—	double bond
three flexible connectors	—	triple bond

A molecular model uses different color balls to represent hydrogen, carbon, oxygen, chlorine, and nitrogen atoms. The color code for each ball is as follows:

white ball	—	hydrogen (one hole)
black ball	—	carbon (four holes)
red ball	—	oxygen (two holes)
green ball	—	chlorine (one hole)
orange ball	—	bromine (one hole)
purple ball	—	iodine (one hole)
blue ball	—	nitrogen (three holes)

Note: If the blue nitrogen ball has more than three holes, use a small peg or tape to fill the additional hole(s). All holes in each ball must have a connector for a model to be built correctly.

PROCEDURE

A. Molecular Models of Hydrocarbons*

Construct the molecular models for the following hydrocarbons and draw their expanded structural formulas in the Data Table.

1. *Alkanes*
   (a) methane, CH_4
   (b) ethane, C_2H_6
   (c) propane, C_3H_8
   (d) butane, C_4H_{10}

2. *Alkenes*
   (a) ethene ("ethylene"), C_2H_4
   (b) propene ("propylene"), C_3H_6

3. *Alkynes*
   (a) ethyne ("acetylene"), C_2H_2
   (b) propyne ("methyl acetylene"), C_3H_4

4. *Arenes*
   (a) methyl benzene, $C_6H_5–CH_3$
   (b) *para*-dimethyl benzene, $C_6H_4–(CH_3)_2$

---

* A chemical name in quotation marks indicates a common name rather than a systematic name.

B. Molecular Models of Hydrocarbon Derivatives

Construct the molecular models for the following hydrocarbon derivatives, and draw their expanded structural formulas in the Data Table.

1. *Organic Halides*
   (a) "methyl chloride," $CH_3–Cl$

   (b) "ethyl iodide," $CH_3CH_2–I$

2. *Alcohols*
   (a) methanol, $CH_3–OH$
       ("methyl alcohol")

   (b) ethanol, $CH_3CH_2–OH$
       ("ethyl alcohol")
   (c) 1-propanol, $CH_3CH_2CH_2–OH$
       ("propyl alcohol")

   (d) 2-propanol, $CH_3CH(OH)CH_3$
       ("isopropyl alcohol")

3. *Phenols*
   (a) phenol, $C_6H_5–OH$

   (b) *ortho*-methyl phenol, $C_6H_4(CH_3)–OH$

4. *Ethers*
   (a) "dimethyl ether," $CH_3–O–CH_3$

   (b) "diethyl ether," $CH_3CH_2–O–CH_2CH_3$

5. *Amines*
   (a) "methyl amine," $CH_3–NH_2$

   (b) "ethyl amine," $CH_3CH_2–NH_2$
   (c) "propyl amine," $CH_3CH_2CH_2–NH_2$ (d) "isopropyl amine," $CH_3CH(NH_2)CH_3$

6. *Aldehydes*
   (a) methanal, HCHO
       ("formaldehyde")

   (b) ethanal, CH_3CHO
       ("acetaldehyde")

7. *Ketones*
   (a) propanone, CH_3COCH_3
       ("acetone" or "dimethyl ketone")

   (b) butanone, $CH_3COCH_2CH_3$
       ("methyl ethyl ketone")

8. *Carboxylic Acids*
   (a) methanoic acid, HCOOH
       ("formic acid")

   (b) ethanoic acid, CH_3COOH
       ("acetic acid")

9. *Esters*
   (a) methyl methanoate, $HCOOCH_3$
       ("methyl formate")

   (b) ethyl ethanoate, $CH_3COOCH_2CH_3$
       ("ethyl acetate")

10. *Amides*
   (a) methanamide, $HCONH_2$
       ("formamide")

   (b) ethanamide, CH_3CONH_2
       ("acetamide")

C. Unknown Molecular Models

The Instructor will provide numbered molecular models of unknown organic compounds. Draw the expanded structural formula in the Data Table and identify the class of compound from the following hydrocarbon derivatives: *organic halide, alcohol, phenol, ether, amine, aldehyde, ketone, carboxylic acid, ester,* and *amide.*

PRELABORATORY ASSIGNMENT*

1. In your own words, define the following terms:

   carbonyl group

   class of compounds

   functional group

   hydrocarbon

   hydrocarbon derivative

   isomers

   organic chemistry

2. Which element is represented by each of the following colored balls?

   (a) black ball   (b)   red ball

   (c) green ball   (d)   blue ball

3. What molecular model parts are used to construct two carbon atoms joined by a single bond?

4. What molecular model parts are used to construct a model of carbon and oxygen atoms joined by a double bond?

5. State the name for each of the following alkyl groups.

   (a) $CH_3—$ (b) $CH_3—CH_2—$

   (c) $CH_3—CH—CH_3$ (d) $CH_3—CH_2—CH_2—$
   $\qquad\quad|$

Answers in Appendix J

6. Draw and label the ortho, meta, and para isomers of dichlorobenzene.

7. State the name of the class of compound for each of the following functional groups.

(a) $-\overset{|}{\underset{|}{C}}-O-\overset{|}{\underset{|}{C}}-$

(b) $-\overset{|}{\underset{|}{C}}-Cl$

(c) $-\overset{|}{\underset{|}{C}}-NH_2$

(d) $-C-\overset{\overset{\displaystyle O}{\|}}{C}-C-$

(e) $-\overset{\overset{\displaystyle O}{\|}}{C}-OH$

(f) $-\overset{\overset{\displaystyle O}{\|}}{C}-H$

(g) $-\overset{\overset{\displaystyle O}{\|}}{C}-NH_2$

(h) $-\overset{|}{\underset{|}{C}}-OH$

(i) C_6H_5-OH

(j) $-\overset{\overset{\displaystyle O}{\|}}{C}-O-C-$

EXPERIMENT 23 NAME ____________________

DATE ____________________ SECTION ____________________

DATA TABLE

A. Molecular Models of Hydrocarbons **Model Kit #________**

1. *Alkanes*
   (a) methane, CH_4
   (b) ethane, C_2H_6
   (c) propane, C_3H_8
   (d) butane, C_4H_{10}

2. *Alkenes*
   (a) ethene (“ethylene”), C_2H_4
   (b) propene (“propylene”), C_3H_6

3. *Alkynes*
   (a) ethyne (“acetylene”), C_2H_2
   (b) propyne (“methyl acetylene”), C_3H_4

4. *Arenes*
   (a) methyl benzene, $C_6H_5–CH_3$
   (b) *para*-dimethyl benzene, $C_6H_4–(CH_3)_2$

B. Molecular Models of Hydrocarbon Derivatives

1. *Organic Halides*
   (a) "methyl chloride," CH_3–Cl

   (b) "ethyl iodide," CH_3CH_2–I

2. *Alcohols*
   (a) methanol, CH_3–OH
   ("methyl alcohol")

   (b) ethanol, CH_3CH_2–OH
   ("ethyl alcohol")

   (c) 1-propanol, $CH_3CH_2CH_2$–OH
   ("propyl alcohol")

   (d) 2-propanol, $CH_3CH(OH)CH_3$
   ("isopropyl alcohol")

3. *Phenols*
   (a) phenol, C_6H_5–OH

   (b) *ortho*-methyl phenol, $C_6H_4(CH_3)$–OH

4. *Ethers*
   (a) "dimethyl ether," CH_3–O–CH_3

   (b) "diethyl ether," CH_3CH_2–O–CH_2CH_3

5. *Amines*
   (a) "methyl amine," CH_3–NH_2

   (b) "ethyl amine," CH_3CH_2–NH_2

   (c) "propyl amine," $CH_3CH_2CH_2$–NH_2

   (d) "isopropyl amine," $CH_3CH(NH_2)CH_3$

6. *Aldehydes*
   (a) methanal, HCHO
   ("formaldehyde")

   (b) ethanal, CH_3CHO
   ("acetaldehyde")

7. *Ketones*
   (a) propanone, CH_3COCH_3
   ("acetone" or "dimethyl ketone")

   (b) butanone, $CH_3COCH_2CH_3$
   ("methyl ethyl ketone")

8. *Carboxylic Acids*
   (a) methanoic acid, HCOOH
   ("formic acid")

   (b) ethanoic acid, CH_3COOH
   ("acetic acid")

9. *Esters*
   (a) methyl methanoate, $HCOOCH_3$
   ("methyl formate")

   (b) ethyl ethanoate, $CH_3COOCH_2CH_3$
   ("ethyl acetate")

10. *Amides*
    (a) methanamide, $HCONH_2$
    ("formamide")

    (b) ethanamide, CH_3CONH_2
    ("acetamide")

C. Unknown Molecular Models

Model Number	Structural Formula	Class of Compound
#1		#1 _____
#2		#2 _____
#3		#3 _____
#4		#4 _____
#5		#5 _____
#6		#6 _____
#7		#7 _____
#8		#8 _____
#9		#9 _____
#10		#10 _____

1. Identify the class of compound (for example, alcohol) corresponding to each of the following general formulas. The symbol **R** represents an alkyl group; the symbol **Ar** represents an aryl group; and **X** represents a halide.)

    (a) R—OH   _____

    (b) R—NH$_2$ _____

    (c) R—O—R'   _____

    (d) R—X   _____

    (e) Ar—OH   _____

               O
               ‖
    (f) R—C—H   _____

               O
               ‖
    (g) R—C—NH$_2$ _____

               O
               ‖
    (h) R—C—OH   _____

               O
               ‖
    (i) R—C—O—R'   _____

               O
               ‖
    (j) R—C—R'   _____

2. Identify five different functional groups in thyroxine (thyroid hormone). The structural formula for thyroxine is shown below and the functional groups are circled.

3. Draw two isomers for the molecular formula C_2H_6O. Identify the class of compound represented by each isomer.

4. Draw two isomers for the molecular formula $C_2H_4O_2$. Identify the class of compound represented by each isomer.

5. (optional) Based on the suffix ending of the following common organic compounds, identify the class of compound (for example, alcohol) for each of the following.

    (a) amphetamine (stimulant)   _____

    (b) citral (lemon fragrance)   _____

    (c) cortisone (adrenal hormone)   _____

    (d) retinol (vitamin A)   _____

    (e) sulfanilamide (antibacterial)   _____

Separation of Food Colors and Amino Acids

OBJECTIVES

- To separate blue, green, red, and yellow food colors by paper chromatography.
- To identify an amino acid in an unknown solution by paper chromatography.
- To develop the lab skill for preparing and developing a paper chromatogram.

DISCUSSION

Chromatography (kro-muh-TOG-ruh-fee) is a method for separating a chemical mixture into its component compounds. The term chromatography is derived from the Greek *chroma,* meaning "color" and *graph,* meaning "record." The Russian botanist Mikhail Tsvett is credited with the first application of the technique, when he separated colored plant pigments in the early 1900s.

Chromatography is a powerful tool that requires only tiny amounts of sample and is routinely employed in **biochemistry** analyses. The technique involves putting a small drop of solution on an adsorbent, such as paper or silica gel, and placing the sample in a chamber with a liquid solvent. In paper chromatography, a drop of sample is placed on a sheet of paper, and the compounds in the mixture separate as the solvent travels up the paper by capillary action.

A compound in a mixture that is more strongly attracted to the solvent moves farther up the paper. A compound less attracted to the solvent moves less, and remains closer to the initial spot. Thus, as the solvent travels up the paper, the compounds in a mixture begin to separate. Food dyes separate into colored, visible spots. However, amino acids separate into colorless, invisible spots, and we must treat the paper to make the spots visible. For example, if we spray the chromatogram with the chemical ninhydrin, amino acid spots change from colorless to purple.

Every compound is characterized by its attraction to the solvent and the adsorbent paper. A retention factor (R_f) indicates the relative attraction of the compound for solvent and for the paper. The **R_f value** expresses the ratio of the distance traveled by the compound compared to the distance traveled by the solvent. The leading edge of the solvent is called the **solvent front**.

Example Exercise 24.1 • Determining an R_f Value

Paper chromatography is used to analyze four biological samples (A, B, C, D), which give the following result using alcohol solvent. State the number of components in each sample, and calculate the R_f value for the compound shown in lane 1.

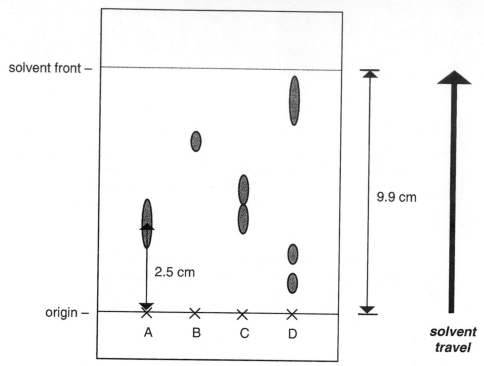

Solution: Samples A and B each have one component as shown by a single spot in lanes 1 and 2. Sample C has two components and D has three components as shown by the spots in lanes 3 and 4.

The distance from the origin to the solvent front is 9.9 cm. The distance from the origin to the center of the spot in lane 1 is 2.5 cm. Therefore, we can calculate the R_f value for the compound in lane 1 as follows.

$$R_f = \frac{2.5 \text{ cm}}{9.9 \text{ cm}} = 0.25$$

We can describe the theory of paper chromatography as follows. Cellulose paper adsorbs moisture from the air and forms a **stationary phase** as the water vapor is very strongly attracted to the paper. After "spotting" a paper chromatogram, it is placed in a developing chamber containing a liquid solvent. The solvent is a **mobile phase** that travels up the paper and across the spot. As the solvent travels across the sample mixture, it begins pulling compounds up the paper. To understand the process, visualize two liquid sheets moving across one another while competing for a compound. This competition—between the mobile phase and the stationary phase—is responsible for separating the components in a mixture.

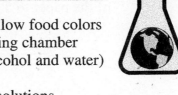

- Whatman No. 1 chromatography paper
- glass capillary tubes for spotting samples
- 1000-mL beaker
- glass stirring rod
- adhesive tape
- graduated cylinder
- long-stem funnel
- aluminum foil, ~15 × 15 cm square
- surgical latex gloves
- 110°C drying oven (or heat lamps)
- metric ruler (e.g., Figure 3.3)

- blue, green, red, yellow food colors
- solvent for developing chamber
  (65:35 isopropyl alcohol and water)

- known amino acid solutions
  glycine solution, 0.1 M Gly
  alanine solution, 0.1 M Ala
  phenylalanine solution, 0.1 M Phe
- unknown amino acid solutions
  0.1 M Gly, 0.1 M Ala, 0.1 M Phe
- 2% ninhydrin in ethanol

PROCEDURE

A. Separation of Food Colors by Paper Chromatography

1. Cut a piece of chromatography paper that measures 9.0 × 15.0 cm. Using a pencil, draw a line 2.0 cm from the bottom of the paper, and mark four equidistant points (×) along the origin line. Label the four points B, G, R, and Y as shown below.

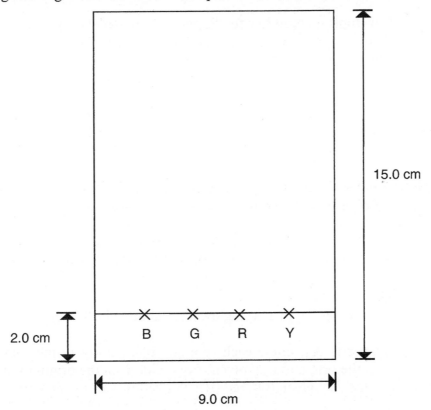

2. Using separate glass capillary tubes, make a 2-mm spot (•) on the origin line at B, G, R, and Y with blue, green, red, and yellow food colors.

   Note: If the initial spot is too large, the components can bleed into an adjacent spot and confuse the results. To avoid a large initial spot, practice spotting, using the capillary tube on a scrap of chromatography paper.

3. Adjust the length of the chromatogram to fit a 1000-mL beaker, and tape the top edge of the paper to a glass stirring rod as illustrated in Figure 24.1.

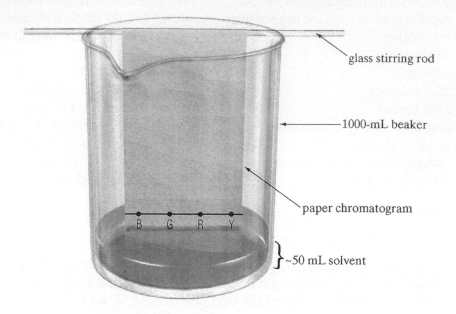

glass stirring rod

1000-mL beaker

paper chromatogram

~50 mL solvent

Figure 24.1 Developing Chamber for Paper Chromatography
The initial solvent level must reach the bottom edge of the paper, but should not touch the sample spots on the origin line.

4. Measure ~50 mL of solvent into a graduated cylinder, and without splashing pour the solvent through a long-stem funnel into the bottom of the 1000-mL beaker. Cover the top of the beaker with a piece of aluminum foil.

   Note: While the food colors are separating in the developing chamber, go on to **Procedure B**, Identification of Amino Acids.

5. After the solvent has moved about halfway up the paper chromatogram (~1.5 hours), remove the paper from the beaker, and draw a dashed line (---) along the solvent front.

6. Place the chromatogram in a drying oven (or under a heat lamp) for ~5 minutes.

7. When the paper is dry, circle each spot and place a dot in the center. Measure the distance from the origin to the solvent front, and from the origin to the center of each spot for the blue (B); green (G); red (R); and yellow (Y) food colors.

   Note: The components in a food color can vary with brand name, but each food color will usually have more than one component.

8. Calculate the R_f value for each component in the blue (B); green (G); red (R); and yellow (Y) food colors.

B. Identification of Amino Acids by Paper Chromatography

1. Put on surgical latex gloves, and cut a rectangular piece of chromatography paper that measures 9.0 × 15.0 cm. Draw a line 2.0 cm from the bottom edge of the paper, and mark four equidistant points (×) along the origin. Label the four points Gly, Ala, Phe, and Unk as shown below.

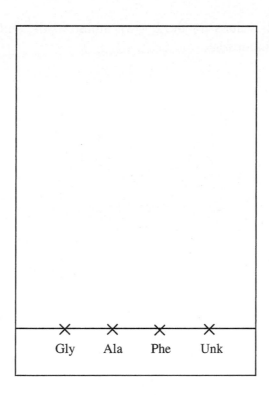

2. Record the unknown number for Unk in the Data Table. Using separate capillary tubes, make a 2-mm spot (•) on the origin line at Gly, Ala, Phe, and Unk with glycine, alanine, phenylalanine, and the unknown amino acid solution.

3. Adjust the length of the chromatogram to fit a 1000-mL beaker, and tape the top edge of the paper to a glass stirring rod as illustrated in Figure 24.1.

4. Measure ~50 mL of solvent into a graduated cylinder, and carefully (without splashing) pour the solvent through a long-stem funnel into the bottom of the 1000-mL beaker. Cover the top of the beaker with a piece of aluminum foil.

5. After the solvent has moved about halfway up the paper chromatogram (~1.5 hours), remove the paper from the beaker, and draw a dashed line (---) along the solvent front.

6. Place the chromatogram in a drying oven (or under a heat lamp) for ~5 minutes.

7. When the paper is dry, spray the chromatogram with ninhydrin solution under a hood (avoid breathing and contact). Dry the chromatogram.

8. Circle each spot on the dry paper with a pencil and place a dot in the center of the spot. Measure the distance from the origin to the solvent front, and from the origin to the center of each spot for Gly, Ala, Phe, and Unk.

9. Calculate the R_f value for glycine (Gly); alanine (Ala); phenylalanine (Phe); and the unknown amino acid (Unk).

10. Refer to the R_f values for the known amino acids, and identify the unknown amino acid as Gly, Ala, or Phe.

EXPERIMENT 24 NAME ______________________

DATE ______________ SECTION ______________________

*PRELABORATORY ASSIGNMENT**

1. In your own words, define the following terms:

   biochemistry

   chromatography

   mobile phase

   R_f value

   solvent front

   stationary phase

2. What is the "origin" on a chromatogram?

3. What is meant by "spotting" a chromatogram?

4. Which is "lane 1" on a chromatogram?

   Which is the "first" component in a given lane?

---

* *Answers in Appendix J*

5. If an amino acid moved 2.1 cm from the origin on a paper chromatogram and the solvent front traveled 7.0 cm, what is the R_f value of the amino acid?

6. Why should a pencil—*not a pen*—be used to mark the origin on a paper chromatogram?

7. What precautions should be observed in this experiment?

NAME _____

SECTION _____

DATA TABLE

A. Separation of Food Colors by Paper Chromatography

Attach the food color paper chromatogram.

distance from origin to solvent front   _____ cm

distance from origin	Blue (B)	Green (G)	Red (R)	Yellow (Y)
to 1st component	_____ cm	_____ cm	_____ cm	_____ cm
to 2nd component	_____ cm	_____ cm	_____ cm	_____ cm

Calculate the R_f value for each component in the food colors (see Example Exercise 24.1).

1st component	_____	_____	_____	_____
2nd component	_____	_____	_____	_____

B. Identification of Amino Acids by Paper Chromatography   **UNKNOWN #** _____

Circle Amino Acid(s) in the Unknown: **Gly   Ala   Phe**

Attach the amino acid paper chromatogram.

distance from origin to solvent front _____ cm

distance from origin	(Gly)	(Ala)	(Phe)	(Unk)
to amino acid	_____ cm	_____ cm	_____ cm	_____ cm

Calculate the R_f value for each amino acid (see Example Exercise 24.1).

| amino acid | _____ | _____ | _____ | _____ |

Identify the amino acid in the unknown (Unk) from its R_f value.

_____

NAME _____

SECTION _____

POSTLABORATORY ASSIGNMENT

1. The artificial sweetener NutraSweet® was treated with acid and analyzed by chromatography. Amino acid solutions of valine (Val), leucine (Leu), phenylalanine (Phe), tryptophan (Trp), tyrosine (Tyr), aspartic acid (Asp), lysine (Lys), and acid hydrolyzed NutraSweet® (NuS) were spotted on a paper chromatogram and gave the following results.

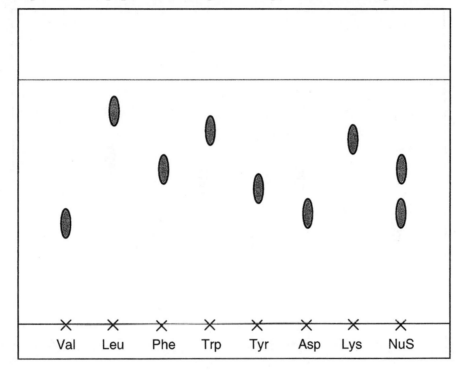

(a) Based on evidence from the chromatogram, how many different amino   _____
acids are released when NutraSweet® is treated with acid? Explain.

(b) Based on evidence from the chromatogram, what are the amino acids   _____
that compose a molecule of NutraSweet®? Explain.

(c) What is the special term for the amide bond in a protein that joins two   _____
amino acids?

(d) What is the special term for a protein molecule that catalyzes (speeds up)   _____
a biochemical reaction?

2. One of the greatest scientific achievements was deciphering the genetic code and understanding how ribonucleic acid (RNA) synthesizes proteins. It was discovered that an RNA codon specifies that a particular amino acid is added to a growing protein chain. A codon is a segment of an RNA molecule composed of three bases; that is, adenine (A), cytosine (C), guanine (G), or uracil (U). The codon GGU is specific for glycine, GCU specifies alanine, and UUU specifies phenylalanine is added to the growing protein chain.

A "stop" codon terminates a protein chain and can be identified by chromatography. Solutions of UGC, UAG, CUA, CGC, AUG, AAA, GUA, and the unknown stop codon, XXX, were spotted and placed in a developing chamber to give the following chromatogram.

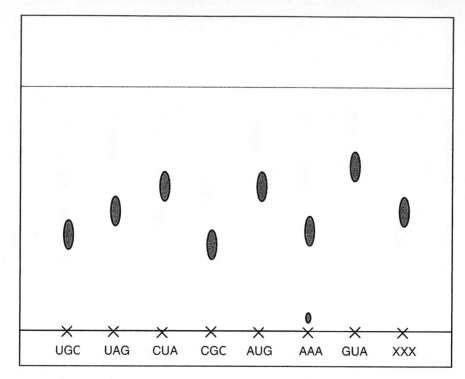

(a) Based on evidence from the chromatogram, what is the identity of the   _____
     "stop" codon? Explain.

(b) Based on evidence from the chromatogram, do any of the solutions   _____
     appear to be contaminated? Explain.

3. (optional) Consider the following questions that refer to paper chromatography.

(a) In theory, can the R_f value be affected by the size of the initial spot?   _____

(b) In practice, can the R_f value be affected by large initial spots, or spots   _____
     very close together? Explain.

Appendices

Laboratory Burner

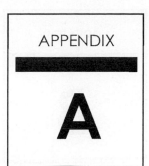

Although a variety of burners are found in chemistry laboratories, they all employ the same principle. Natural gas is allowed to flow into the barrel of the burner and mix with the air, which contains oxygen. The ratio of gas to air can be adjusted, which in turn regulates the temperature of the flame. The more air that is available, the hotter the flame. Two typical burners are shown in Figure A.1.

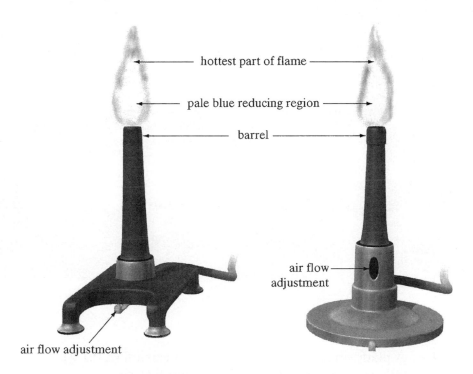

hottest part of flame

pale blue reducing region

barrel

air flow adjustment

air flow adjustment

Figure A.1 Laboratory Burners

Steps in Operating a Burner

1. Close the air flow adjustment.
2. Open the gas jet.
3. Light the burner at the top of the barrel.
4. To obtain a hotter flame, open the air flow adjustment.
5. To shut off the burner, close the gas jet.

Platform Balance

A platform balance provides measurements with an uncertainty of one decigram ($\pm$ 0.1 g). The mass of a sample is determined by placing it on the balance pan, and sequentially adjusting the metal riders on the beams. The heaviest rider (**100-g**) is adjusted first, and the **10-g** rider second. The **1-g** rider is adjusted last, and the balance beam is read to the nearest subdivision (0.1-g). The mass of the sample is equal to the sum of the masses indicated by all of the riders; for example, 100 g + 20 g + 7.2 g = 127.2 g.

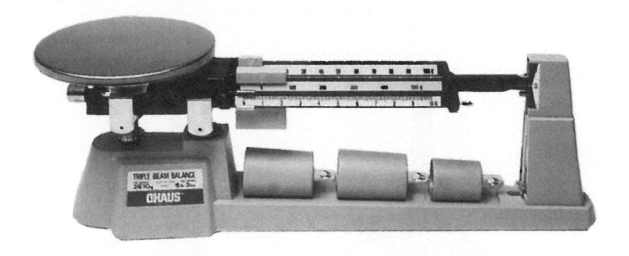

Figure B.1   A platform balance with decigram precision ($\pm$ 0.1 g).
(Photo courtesy of Ohaus Scale Corporation)

Beam Balance

A beam balance provides measurements with an uncertainty of one centigram ($\pm$ 0.01 g). The mass of a sample is determined by placing it on the hanging pan, and sequentially adjusting the metal riders on the beams. The heaviest rider (**100-g**) is adjusted first, the **10-g** rider second, and the **1-g** rider third. The **0.1-g** rider is adjusted last, and the beam is read to the nearest subdivision (0.01-g). The mass of the sample is equal to the sum of the masses indicated by all of the riders; for example, 100 g + 20 g + 7 g + 0.25 g = 127.25 g.

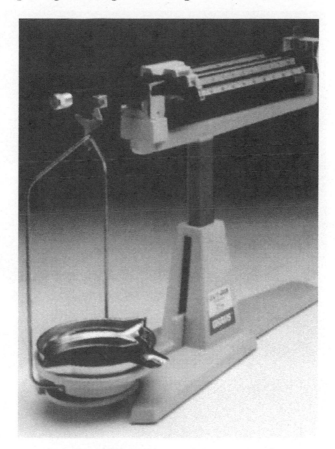

Figure C.1   A beam balance with centigram precision ($\pm$ 0.01 g).
(Photo courtesy of Ohaus Scale Corporation)

Electronic Balance

An electronic balance is more sophisticated than a platform or beam balance. An electronic balance is an expensive instrument that must be used carefully, and it is advisable to have instruction in its operation before using. There are a variety of electronic balance models available. Some balances have milligram uncertainty (± **0.001 g**), some have tenth milligram uncertainty (± **0.0001 g**), and others have hundredth milligram uncertainty (± **0.00001 g**). Figure D.1 illustrates a digital electronic balance.

Figure D.1   A digital electronic balance with milligram precision (± 0.001 g).
            (Photo courtesy of Sartorius Balance Corporation)

Volumetric Pipet

When using a **volumetric pipet**, follow these three steps.

1. Condition the pipet with a small portion of the solution to be transferred.
2. Use a pipet bulb to draw the solution above the calibration line as shown in Figure E.1. Slip the pipet bulb off and place your finger over the end of the pipet. Move your finger slightly so as to allow the bottom of the meniscus to drop slowly to the calibration line.
3. Place the tip of the pipet into a flask or beaker. Allow the pipet to drain free and touch off the last drop on the pipet tip (do not blow out the last drop of solution).

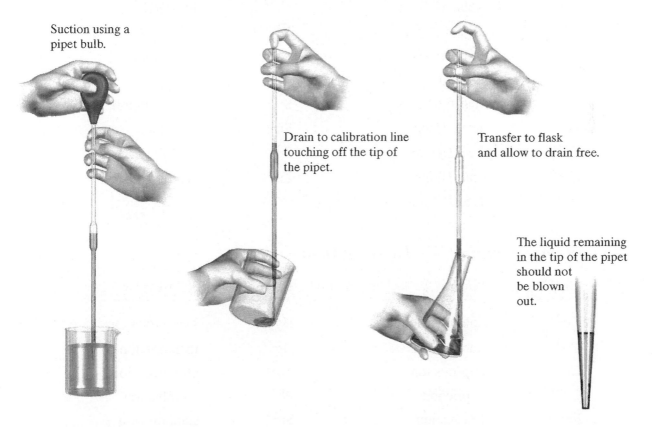

Suction using a pipet bulb.

Drain to calibration line touching off the tip of the pipet.

Transfer to flask and allow to drain free.

The liquid remaining in the tip of the pipet should not be blown out.

Figure E.1 Transferring a quantity of solution using a volumetric pipet.

APPENDIX

F

Common Cations and Anions

Monoatomic Cations — Stock System

Cation	Name	Cation	Name
Al^{3+}	aluminum ion	Li^+	lithium ion
Ba^{2+}	barium ion	Mg^{2+}	magnesium ion
Cd^{2+}	cadmium ion	Mn^{2+}	manganese(II) ion
Ca^{2+}	calcium ion	Hg_2^{2+}	mercury(I) ion
Co^{2+}	cobalt(II) ion	Hg^{2+}	mercury(II) ion
Co^{3+}	cobalt(III) ion	Ni^{2+}	nickel(II) ion
Cu^+	copper(I) ion	K^+	potassium ion
Cu^{2+}	copper(II) ion	Ag^+	silver ion
Cr^{3+}	chromium(III) ion	Na^+	sodium ion
H^+	hydrogen ion	Sr^{2+}	strontium ion
Fe^{2+}	iron(II) ion	Sn^{2+}	tin(II) ion
Fe^{3+}	iron(III) ion	Sn^{4+}	tin(IV) ion
Pb^{2+}	lead(II) ion	Zn^{2+}	zinc ion
Pb^{4+}	lead(IV) ion		

Monoatomic Cations — Latin System

Cation	Name	Cation	Name
Co^{2+}	cobaltous ion	Hg_2^{2+}	mercurous ion
Co^{3+}	cobaltic ion	Hg^{2+}	mercuric ion
Cu^+	cuprous ion	Pb^{2+}	plumbous ion
Cu^{2+}	cupric ion	Pb^{4+}	plumbic ion
Fe^{2+}	ferrous ion	Sn^{2+}	stannous ion
Fe^{3+}	ferric ion	Sn^{4+}	stannic ion

Monoatomic Anions

Anion	Name	Anion	Name
Br^-	bromide ion	N^{3-}	nitride ion
Cl^-	chloride ion	O^{2-}	oxide ion
F^-	fluoride ion	P^{3-}	phosphide ion
I^-	iodide ion	S^{2-}	sulfide ion

Polyatomic Cations

Cation	Name	Cation	Name
NH_4^+	ammonium ion	H_3O^+	hydronium ion

Polyatomic Anions

Anion	Name	Anion	Name
$C_2H_3O_2^-$	acetate ion	ClO^-	hypochlorite ion
CO_3^{2-}	carbonate ion	NO_3^-	nitrate ion
ClO_3^-	chlorate ion	NO_2^-	nitrite ion
ClO_2^-	chlorite ion	$C_2O_4^{2-}$	oxalate ion
CrO_4^{2-}	chromate ion	ClO_4^-	perchlorate ion
CN^-	cyanide ion	MnO_4^-	permanganate ion
$Cr_2O_7^{2-}$	dichromate ion	PO_4^{3-}	phosphate ion
HCO_3^-	hydrogen carbonate ion	SO_4^{2-}	sulfate ion
HSO_4^-	hydrogen sulfate ion	SO_3^{2-}	sulfite ion
OH^-	hydroxide ion	$S_2O_3^{2-}$	thiosulfite ion

Activity Series

Li
K
Ba
Sr
Ca
Na
Mg
Al
Mn
Zn
Fe
Cd
Co
Ni
Sn
Pb

(H)

Cu
Ag
Hg
Au

APPENDIX

H

Solubility Rules for Ionic Compounds

Compounds containing the following ions are generally soluble in water:

1. alkali metal ions and ammonium ions, Li^+, Na^+, K^+, NH_4^+

2. acetate ion, $C_2H_3O_2^-$

3. nitrate ion, NO_3^-

4. halide ions (X), Cl^-, Br^-, I^-
   (AgX, Hg_2X_2, and PbX_2 are exceptions and *insoluble*)

5. sulfate ion, SO_4^{2-}
   ($SrSO_4$, $BaSO_4$, and $PbSO_4$ are exceptions and *insoluble*)

Compounds containing the following ions are generally insoluble* in water:

6. carbonate ion, CO_3^{2-}
   (see Rule 1 exceptions, which are *soluble*)

7. chromate ion, CrO_4^{2-}
   (see Rule 1 exceptions, which are *soluble*)

8. phosphate ion, PO_4^{3-}
   (see Rule 1 exceptions, which are *soluble*)

9. sulfide ion, S^{2-}
   (CaS, SrS, BaS, and Rule 1 exceptions are *soluble*)

10. hydroxide ion, OH^-
    [$Ca(OH)_2$, $Sr(OH)_2$, $Ba(OH)_2$, and Rule 1 exceptions are *soluble*]

---

* These compounds are actually slightly soluble, or very slightly soluble, in water.

APPENDIX

Glossary

–A–

abscissa The horizontal axis (x-axis) on a graph.

activity series A relative order of metals arranged in a list according to their ability to undergo reaction. A metal higher in the series will displace another metal from its aqueous solution.

actual yield The amount of product experimentally obtained from a reaction.

alkali metal Any Group IA/1 element in the periodic table, excluding hydrogen.

alkaline earth metal Any Group IIA/2 element in the periodic table.

anhydrous A term for a substance that does not contain water.

anion Any negatively charged ion.

aqueous solution A solution of a substance dissolved in water.

atmospheric pressure The pressure exerted by the air molecules in Earth's atmosphere; atmospheric pressure is measured with a barometer.

"atomic fingerprint" A term for the unique line spectrum that is characteristic of a given element and can be used for identification.

Avogadro's number (N) The value that corresponds to the number of carbon atoms in 12.01 g of carbon; 6.02×10^{23} particles.

–B–

Balmer formula A mathematical equation for calculating the emitted wavelength of light from an excited hydrogen atom when an electron drops to the second energy level.

biochemistry The study of compounds derived from plants and animals.

-C-

carbonyl group The C=O group, which is present in aldehydes, ketones, carboxylic acids, esters, and amides.

catalyst A substance that speeds up a chemical reaction and is usually recovered without change after the reaction is complete.

cation Any positively charged ion.

centimeter (cm) A common metric unit of length; there are 100 cm in a meter.

centrifuge An instrument that uses centrifugal force to separate a precipitate from solution. The act of rapidly spinning a test tube to isolate precipitate particles at the bottom of a test tube.

change of state The conversion from one physical state to another; for example, the change in a substance from a liquid to a solid.

chemical change A process whereby a substance undergoes a change in chemical formula or chemical composition.

chemical property A characteristic of a substance that cannot be observed without changing the chemical composition of the substance.

chemistry The branch of science that studies the composition and properties of matter.

chromatography A method for separating a mixture into its components as a result of a varying attraction of compounds for a mobile solvent on a stationary solid.

class of compounds A family of compounds in which all the members have the same structural feature (that is, an atom or group of atoms) and similar chemical properties.

combined gas law The pressure exerted by a gas is inversely proportional to its volume and directly proportional to its Kelvin temperature.

compound A pure substance that can be broken down into two or more simpler substances by chemical reaction.

conditioning Rinsing a piece of glassware with a sample liquid or solution to prevent dilution by water on the inside surface; pipets and burets are usually conditioned before use.

continuous spectrum A broad uninterrupted band of radiant energy.

covalent bond A bond characterized by the sharing of one or more pairs of valence electrons.

-D-

Dalton's law of partial pressures The pressure exerted by a mixture of gases is equal to the sum of the pressures exerted by each gas in the mixture.

decant The process of pouring a liquid from one container into another; for example, pouring the supernate from one test tube into a second test tube.

density (*d*) The amount of mass in a unit volume of matter.

digestion The process of heating a precipitate in solution in order to develop larger particles that are easier to filter and free of impurities.

dissociation The process of an ionic compound dissolving in water and separating into positive and negative ions.

double bond A bond between two atoms composed of two electron pairs and shown as two dashes between the symbols of two atoms.

–E–

electron dot formula A diagram of a molecule in which each atom is represented by its chemical symbol surrounded by two dots for each pair of bonding or nonbonding electrons.

element A pure substance that cannot be broken down any further by ordinary chemical reaction.

empirical formula The chemical formula of a compound that expresses the simplest whole number ratio of atoms of each element in a molecule, or ions in a formula unit.

endpoint The stage in a titration when the indicator changes color.

experimental conditions The conditions of temperature and pressure at which a gas sample is collected; not usually STP.

experiment A scientific procedure for collecting data and recording observations under controlled conditions.

–F–

filtrate The solution that passes through the filter paper in a filtration operation.

firing to red heat Heating a crucible or other porcelain object until it glows red.

flame test A test for an element's characteristic color by placing a sample in a hot flame.

freezing point The temperature at which a liquid substance crystallizes and forms a solid.

frequency The number of times a light wave travels a complete cycle in one second.

functional group An atom or group of atoms that characterizes a class of compounds, and contributes to their similar physical and chemical properties.

–G–

gram (**g**) The basic unit of mass in the metric system; 1000 g equal a kilogram.

group A vertical column in the periodic table; a family of elements with similar properties.

–H–

halide A term referring to any negatively charged Group VIIA/17 atom; for example, bromide, chloride, or iodide.

halogen Any Group VIIA/17 element; that is, fluorine, chlorine, bromine, or iodine.

heating to constant weight A repeated process of heating, cooling, and weighing until the mass reading for an object are constant, or agree closely.

heterogeneous mixture Matter having an indefinite composition and properties that can vary within the sample.

homogeneous mixture Matter having a definite composition but properties that can vary from sample to sample; examples include alloys, solutions, and gas mixtures.

hydrate A substance that contains a specific number of water molecules attached to a formula unit in a crystalline compound.

hydrocarbon A compound containing only hydrogen and carbon.

hydrocarbon derivative A compound containing carbon, hydrogen, and another element such as oxygen, nitrogen, or a halogen.

hypothesis A tentative proposal of a scientific principle that attempts to explain the meaning of a set of data collected in an experiment.

–I–

immiscible A term that refers to liquids that do not dissolve in one another and separate into two layers.

indicator A chemical substance that undergoes a color change according to the pH of a solution; for example, phenolphthalein is colorless below pH 9 and pink above pH 9.

ionization The process of a polar molecular compound dissolving in water and forming positive and negative ions.

isomers Compounds with the same molecular formula but with different structural formulas. Isomers have different physical and chemical properties.

–L–

light A specific term that refers to the portion of the radiant energy spectrum that is visible; that is, violet through red. A general term that refers to all forms of radiant energy.

***like dissolves like* rule** The general principle that solubility is greatest when the polarity of the solute is similar to that of the solvent.

line spectrum The narrow bands of light observed through a spectroscope that are emitted from excited atoms in a gas discharge tube.

–M–

mass The amount of matter in a sample. Mass is independent of Earth's gravitational attraction and is the quantity measured with a laboratory balance.

mass/mass percent concentration (m/m %) A solution concentration expression that relates the mass of solute in grams dissolved in each 100 grams of solution.

$$\frac{\text{mass of solute}}{\text{mass of solution}} \times 100\% = \text{m/m \%}$$

melting point The temperature at which a solid substance melts and forms a liquid.

meniscus The lens-shaped appearance at the surface of a liquid inside a piece of narrow glassware, such as a graduated cylinder, pipet, or buret.

metric system A decimal system of measurement using prefixes and a basic unit to express physical quantities such as length, mass, and volume.

milliliter (mL) A common metric unit of volume; there are 1000 mL in a liter.

miscible A term that refers to liquids that dissolve completely in one another.

mobile phase A term that refers to the solvent that travels across a paper chromatogram by capillary action.

molar concentration (M) A solution concentration expression that relates the moles of solute dissolved in each liter of solution; also referred to as molarity.

$$\frac{\text{moles of solute}}{\text{liters of solution}} = M$$

molar mass (MM) The mass of 1 mole of any substance expressed in grams. (The individual particles that compose the substance may be atoms, molecules, or formula units.)

molar volume The volume occupied by 1 mole of any gas at STP; at 0°C and 760 mm Hg, the volume of 1 mole of any gas is 22.4 L (22,400 mL).

mole (mol) The amount of substance that contains Avogadro's number of particles; that is, an amount of substance that contains 6.02×10^{23} particles.

molecular formula The chemical formula of a compound that expresses the actual number of atoms of each element in a molecule.

monolayer A term that refers to a film layer of organic molecules on the surface of water; the monolayer is only 1 molecule thick.

–N–

nanometer (nm) A unit of length used to express wavelengths of light; a unit of length equal to one-billionth of a meter.

net ionic equation A chemical equation that portrays an ionic reaction after spectator ions have been canceled from the total ionic equation. The net ionic equation shows only those species undergoing a change during a chemical reaction.

–O–

octet rule The statement that an atom tends to bond in such a way so as to acquire eight electrons in its outer shell. A hydrogen atom is an exception to the rule and acquires only two valence electrons.

ordinate The vertical axis (y-axis) on a graph.

organic chemistry The study of carbon compounds.

origin The point of intersection of the horizontal and vertical axes on a graph.

oxidation A process in which a substance undergoes an increase in oxidation number. A process characterized by losing electrons.

oxidation number A positive or negative value assigned to an atom in a substance according to a set of rules. A value that indicates whether an atom is electron poor or electron rich compared to a free atom. Metals and nonmetals in the free state have an oxidation number of zero.

oxidizing agent A substance that causes the oxidation of another substance in a redox reaction. The substance that is reduced in a redox reaction.

–P–

percent yield An expression for the amount of the actual yield compared to the theoretical yield.

period A horizontal row in the periodic table; a series of elements with properties that vary from metallic to nonmetallic.

periodic law The properties of the elements recur in a repeating pattern when arranged according to increasing atomic number.

photon A particle of light that corresponds to a unit of radiant energy. A photon may also be referred to as a quantum (*pl.,* quanta).

physical change The process of undergoing a change without altering the chemical formula of a substance; for example, a change in physical state.

physical property A characteristic that can be observed without changing the chemical composition of the substance.

physical state A term for the condition of a substance existing as a solid, liquid, or gas.

precipitate (ppt) An insoluble solid substance produced from a reaction in aqueous solution.

product A substance resulting from a chemical reaction.

–Q–

qualitative analysis A systematic procedure for the separation and identification of cations, anions, or other substances present in a sample.

–R–

reactant A substance undergoing a chemical reaction.

redox reaction A chemical reaction that involves electron transfer between two reacting substances and causes reduction of one species and oxidation of another.

reducing agent A substance that causes the reduction of another substance in a redox reaction. The substance that is oxidized in a redox reaction.

reduction A process in which a substance undergoes a decrease in oxidation number. A process characterized by gaining electrons.

R_f value The ratio of the distance traveled by a sample component compared to the distance traveled by the solvent.

–S–

science The methodical exploration of nature and the logical explanation of the observations.

scientific method A systematic investigation that involves performing an experiment, proposing a hypothesis, testing the hypothesis, and stating a theory or law that explains a scientific principle.

single bond A bond between two atoms composed of one electron pair. A single bond is represented as a dash between the symbols of two atoms.

solute The component of a solution that is the lesser quantity.

solvent The component of a solution that is the greater quantity.

solvent front The leading edge of the solvent, which travels from the bottom of the developing chamber to the upper portion of the chromatogram.

spectator ions Those ions that are in aqueous solution, but do not participate in a reaction or appear as reactants or products in the net ionic equation.

standard conditions See standard temperature and pressure.

standardization A procedure for establishing the concentration of a solution precisely, usually to three or four significant digits.

standard temperature and pressure (STP) A temperature of 0°C and a pressure of 1 atm. A temperature of 273 K and a pressure of 760 mm Hg for a gas.

stationary phase A term for the moisture that is strongly adsorbed onto the paper chromatogram and is not free to travel.

stoichiometry The relationship of quantities (mass of substance or volume of gas) in a chemical change according to the balanced chemical equation.

strong electrolyte An aqueous solution that is a good conductor of electricity and produces a bright glow in a standard light bulb in a conductivity apparatus.

structural formula A diagram of a molecule or polyatomic ion that shows the chemical symbol of each atom and a dash representing each pair of bonding electrons.

sublimation The direct change of state from a solid to a gas without forming a liquid. Conversely, the direct change of state from a gas to a solid is called deposition.

substance Matter having constant composition with definite and predictable properties.

supernate The liquid in contact with a precipitate after insoluble particles have been centrifuged or settled from solution.

supersaturated solution A solution containing more solute than can ordinarily dissolve at a given temperature. A supersaturated solution is unstable and the excess solute will crystallize from solution if a seed crystal is added.

surface area The region occupied by the layer of organic molecules floating on water; the formula for calculating surface area is $\pi d^2/4$.

–T–

theoretical yield The amount of product that is calculated from given amounts of reactants.

theory An extensively tested proposal of a scientific principle that explains the behavior of nature. A theory offers a model, for example the atomic theory, to describe nature.

titration A laboratory procedure for delivering a measured volume of solution through a buret.

total ionic equation A chemical equation that writes highly ionized substances in the ionic form and slightly ionized substances in the nonionized form.

triple bond A bond between two atoms composed of three electron pairs and represented by three dashes between the symbols of the two atoms.

–U–

uncertainty A term that refers to the degree of inexactness in an instrumental measurement; for example, ± 0.05 cm, ± 0.001 g, ± 0.5 mL, $\pm 0.5°C$, or ± 1 s.

–V–

valence electrons The electrons in the outermost s and p energy sublevels that are available for chemical bonding.

vapor pressure The pressure exerted by gaseous vapor above a liquid in a closed container when the rates of evaporation and condensation are equal; for example, the pressure exerted by water vapor above liquid water.

visible spectrum Light energy that is observed as violet, blue, green, yellow, orange, and red; the region in the radiant energy spectrum from approximately 400–700 nm.

volume by displacement A technique for determining the volume of a sample by measuring the volume of water it displaces.

–W–

water of crystallization A term for the number of water molecules bound to a formula unit in a hydrate; also called *water of hydration.*

wavelength (λ) The distance a light wave travels to complete one cycle.

weak electrolyte An aqueous solution that is a poor conductor of electricity and produces a dim glow in a standard light bulb in a conductivity apparatus.

weighing by difference A procedure for obtaining the mass of a sample indirectly by first weighing a container and then weighing the container with the sample. Chemical samples are not usually weighed directly, as they could corrode the balance.

Answers to Prelaboratory Assignments

Experiment 1 — *Introduction to Chemistry*

1. See the Glossary, Appendix I.

2. Refer to the diagrams of Common Laboratory Equipment, pages 4–5.

3. All chemicals—*even water*—can be dangerous, as you will observe in the experiment.

4. Flush immediately with water and notify the Instructor of any irritation.

5. • Wear eye protection; be careful when handling chemicals and solutions.

   • Handle glassware carefully, as it is easily broken and can cause cuts.

Experiment 2 — *Instrumental Measurements*

1. See the Glossary, Appendix I.

2. (a) ± 0.1 cm; (b) ± 0.05 cm; (c) ± 0.1 g; (d) ± 0.01 g; (e) ± 0.001 g; (f) ± 0.5 mL; (g) ± 0.5 °C

3. 8.8 cm, 14.0 cm, 7.15 cm; 10.00 cm

4. 14.0 mL, 83.5 mL

5. 30.0 °C, 2.5 °C

6. $15.6 = 16$ cm^3

7. $15.3125 = 15.3$ cm^3

8. • Wear eye protection; be careful when using the laboratory burner.

   • Handle the laboratory burner and boiling waterbath carefully, as they can cause burns.

   • Handle the thermometer carefully, as it is easily broken and can cause cuts.

   (Report a broken thermometer immediately to the Instructor; mercury vapor is poisonous.)

Experiment 3 — Density of Liquids and Solids

1. See the Glossary, Appendix I.
2. 54.0 mL, 62.5 mL
3. 5.40 cm, 4.25 cm
4. 0.790 g/mL
5. 1.2 g/mL
6. 4.51 g/cm^3
7. 0.00185 cm (1.85 × 10^{-3} cm)
8. • Wear eye protection; be careful when handling the glass pipet.
    • Keep the flammable unknown liquids and vapors away from a laboratory burner flame.
    • Dispose of the unknown liquids in an organic waste container designated by the Instructor.

Experiment 4 — Freezing Points and Melting Points

1. See the Glossary, Appendix I.
2. Hot tap water can leave mineral deposits on glassware.
3. ~40 °C
4. above 65 °C
5. 65.0 °C
6. The freezing point corresponds to the flat plateau portion of the curve. The freezing point is determined by extending a straight line from the plateau on the curve back to the vertical axis.
7. Place the test tube containing the solid paradichlorobenzene in hot water. After the compound melts, remove the thermometer and wipe off any residue with a paper towel.
8. Determining a melting point is time-consuming if the substance is heated 1 °C per minute. Therefore, heat the water rapidly for the first trial, and determine an approximate melting point. In the second trial, heat the water rapidly to within a few degrees of the approximate melting point and then slowly (about 1 °C per minute) to determine an accurate melting point.
9. 65–75 °C
10. 68.0–69.5 °C
11. If the compound appears to liquefy, the problem may be that
    (1) The temperature of the waterbath is higher than the melting point of the compound.
    (2) The capillary may not be sealed completely, and water may be leaking into the tube.
12. • Wear eye protection; be careful when using the laboratory burner.
    • Do not heat the test tube directly because paradichlorobenzene is flammable.
    • Do not pour out the liquid paradichlorobenzene, as it is used for repeated trials.
    • Handle the thermometer carefully, and do not attempt to remove a thermometer frozen in solid paradichlorobenzene. (Report a broken thermometer immediately to the Instructor.)

Experiment 5 — *Physical Changes and Chemical Changes*

1. See the Glossary, Appendix I.
2. (a) phys; (b) phys; (c) phys; (d) phys; (e) phys; (f) phys; (g) phys; (h) phys; (i) phys; (j) chem
3. (a) phys; (b) phys; (c) chem; (d) phys; (e) phys; (f) chem; (g) chem; (h) chem; (i) chem; (j) chem
4. The boiling chip prevents "bumping," which ejects flammable liquid from the test tube.
5. All of the following suggest a chemical change:

   (1) a solution releases gas bubbles; (2) a solution forms an insoluble solid; (3) a solution undergoes a permanent color change; (4) a solution releases or absorbs energy.
6. A gas is released if there is fizzing or bubbling, or an odor is observed.
7. • Wear eye protection; be careful when determining a boiling point and keep the flammable liquids and vapors away from a burner flame.

   • Dispose of the liquids in an organic waste container designated by the Instructor.

   • Avoid breathing iodine vapors and keep the beaker covered with an evaporating dish while heating the I_2 crystals. (Place only 3 small iodine crystals in the beaker.)

   • Handle the thermometer carefully, as it is easily broken and can cause cuts.

   (Report a broken thermometer immediately to the Instructor; mercury vapor is poisonous.)

Experiment 6 — *"Atomic Fingerprints"*

1. See the Glossary, Appendix I.
2. (a) decreasing wavelength: $R > O > Y > G > B > V$ (Violet has the shortest wavelength.)

   (b) decreasing frequency: $V > B > G > Y > O > R$ (Red has the lowest frequency.)

   (c) decreasing energy: $V > B > G > Y > O > R$ (Red has the lowest energy.)
3. 450 nm, 570 nm, 630 nm
4. violet, blue-green, red
5. We find the wavelength of light emitted when electrons drop from the 10th to 2nd energy level

   using the Balmer formula:
   $$\frac{1}{\lambda} = \frac{1}{91 \text{ nm}} \left(\frac{1}{2^2} - \frac{1}{10^2} \right) = \frac{0.24}{91 \text{ nm}}$$

   and taking the reciprocal:
   $$\lambda = \frac{91 \text{ nm}}{0.24} = 379 \text{ nm} = 380 \text{ nm}$$

   Since 91 nm has two significant digits, the calculation rounds to two significant digits.
6. When 1 electron drops from the 10th to 2nd energy level (or any level), 1 photon of light is emitted. If 5 electrons drop from the 10th to 2nd energy level, 5 photons are emitted.
7. • Be careful not to drop the hand spectroscope, as the wavelength scale can misalign.

   • To avoid a burn, do not touch a hot gas discharge tube in the power supply.

   • To avoid a shock, do not touch the discharge tube while the power supply is turned on.

Experiment 7 — *Families of Elements*

1. See the Glossary, Appendix I.

2. lithium, potassium, sodium; barium, calcium, strontium; bromide, chloride, iodide

3. Sodium contamination is ever-present and may give a false, yellow flame test for sodium.

4. A brief green flame test indicates the presence of barium; a brick-red flame indicates calcium; a scarlet-red flame indicates lithium; a violet flame indicates potassium; a strong orange-yellow flame indicates sodium; and a bright red flame indicates strontium.

5. Water and hexane are *immiscible* and separate into two layers. Hexane is less dense than water and is found in the upper layer. The halide test is observed in the upper hexane layer.

6. • Wear eye protection; be careful when using the burner and performing a flame test.
   • Handle acids carefully, and avoid breathing the vapors of concentrated hydrochloric acid.
   • Dispose of hexane liquid in an organic waste container designated by the Instructor.

Experiment 8 — *Identifying Cations in Solution*

1. See the Glossary, Appendix I.

2. Tap water contains many ions, some of which can interfere with the results of the analysis.

3. A brief green flame test in test tube #1 confirms Ba^{2+}; a brick-red flame test in test tube #2 confirms Ca^{2+}; and a blue gel precipitate in test tube #3 confirms Mg^{2+}.

4. Place a stirring rod into test tube #3 and touch the rod to litmus paper. A blue spot shows the solution is basic. If the spot is red, add a few more drops of NaOH.

5. Refer to **Figure 8.4** to determine the two cations present, and the one cation absent.

6. • Wear eye protection; be careful when using the burner and performing a flame test.
   • Avoid contact with HCl and NaOH. If contacted, wash the area immediately with water.
   • Balance the centrifuge before operating.

Experiment 9 — *Identifying Anions in Solution*

1. See the Glossary, Appendix I.

2. Tap water contains many ions, some of which can interfere with the results of the analysis.

3. A yellow precipitate in test tube #1 confirms I^-. (If the precipitate is white, add 10 drops of water and mix with a stir rod.) A white precipitate in test tube #2 confirms Cl^-; and a white precipitate in test tube #3 confirms SO_4^{2-}. (If the precipitate in test tube #2 or #3 is yellow, it is due to AgI contamination from test tube #1.)

4. Place a stirring rod into test tube #2 and touch the rod to litmus paper. A red spot shows the solution is acidic. If the spot is blue, add a few more drops of HNO_3.

5. Refer to **Figure 9.3** to determine the two anions present, and the one anion absent.

6. • Wear eye protection; avoid contact with HNO_3, NH_4OH, and $AgNO_3$. If contacted, wash the area immediately with water.
   • Balance the centrifuge before operating.

Experiment 10 — Analysis of a Penny

1. See the Glossary, Appendix I.
2. ($\rightarrow$)—gives, produces, or yields; (+)—added to or reacts with; (Δ)—heat; *NR*—no reaction;
   (g)—gas; (l)—liquid state; (s)—solid state or precipitate; (aq)—aqueous solution.
3. (a) a gas is produced; (b) a precipitate is formed; (c) a color change is observed;
   (d) an energy change, such as heat or light, is noted.
4. (a) colorless; (b) pink
5. The acceptable range of mint dates for a "zinc penny" is 1983–present.
6. (a) 2.5% Cu; (b) 97.49% Zn
7. • Wear eye protection; be careful when using the laboratory burner.
   • Igniting magnesium metal ribbon releases intense heat and sparks.
   • Heating sulfur powder in air produces a pungent, unpleasant smelling gas.
   • Handle acids carefully, and avoid breathing the vapors of hydrochloric acid.

Experiment 11 — Determination of Avogadro's Number

1. See the Glossary, Appendix I.
2. –COOH.
3 The "contact lens" drop disappears.
4. When the "contact lens" drop remains for 30 seconds without evaporating.
5. • Error is introduced by delivering drops of solution that are too large or vary in size.
   • It is very important that the surface of the watchglass is clean. Touching the watchglass
     surface can leave a trace of skin oil that can lead to meaningless results.
6. (a) 123 cm^2; (b) 5.9×10^{16} molecules; (c) 6.7×10^{-8} mol; (d) 8.8×10^{23} molecules/mol
7. • Wear eye protection; be careful with the dropper pipet, as it is sharp and delicate.
   • Avoid breathing the organic hexane vapor.
   • Keep the flammable hexane liquid and vapor away from a laboratory burner flame.
   • Dispose of excess stearic acid solution in a waste container designated by the Instructor.

Experiment 12 — Empirical Formulas of Compounds

1. See the Glossary, Appendix I.
2. The empty crucible and cover are fired to red heat in order to burn off impurities in the crucible and to establish a constant weight.

3. The suggested periods for heating and cooling are general guidelines. More important, the metal should be heated long enough for complete conversion to the oxide. The crucible should be cooled sufficiently to avoid a weighing error.

4. Igniting magnesium in air produces magnesium nitride in addition to the magnesium oxide. Adding distilled water to the crucible contents decomposes magnesium nitride and releases ammonia gas. (The crucible contents fizz, and the odor of ammonia gas may be noticeable.) Reheating the crucible and contents converts magnesium hydroxide to magnesium oxide.

5. If the magnesium metal has reacted completely, there will be no small sparks observed when the crucible cover is lifted.

6. We will assume that the copper wire has reacted completely when there are no longer any traces of yellow (or yellow-brown) sulfur in the crucible. If there is any doubt, heat the crucible to constant weight.

7. $Ca_{0.00624}O_{0.00625}$ = CaO

8. • A hot crucible on the balance causes a weighing error, and mass readings will be low.
   • Smoke from the crucible indicates loss of magnesium oxide, and mass readings will be low.
   • Sulfur has a tendency to "creep" out of the crucible during the firing. This excess sulfur must be heated and driven off as a gas, or the mass readings will be high.

9. • Wear eye protection; be very careful when lifting the crucible cover to check the progress of the reaction as sparks may escape from the crucible.
   • Before firing to red heat, set the crucible on the lab bench and strike sharply with a pencil. A crucible with a hairline crack gives a dull ring.
   • A crucible that glows red has a temperature near 1100 °C. Below this temperature, a crucible may not glow red, but it can cause a painful burn. Handle the crucible cover with tongs.
   • The ignition of magnesium is strongly exothermic and frequently cracks crucibles. A few porcelain chips in the bottom of the crucible under the magnesium minimizes this problem.
   • Heating copper and sulfur together produces toxic sulfur dioxide gas. Avoid breathing the gas, and vent the reaction under a fume hood.

Experiment 13 — *Analysis of Alum*

1. See the Glossary, Appendix I.

2. Microwave popcorn pops as moisture in the corn kernel is heated to steam, splits the kernel, and escapes. Thus, the unpopped kernels *weigh more* than the popped corn. Similarly, the mass of the hydrate salt in this experiment *weighs more* than the mass of the anhydrous salt, as steam escapes and collects on the watchglass covering the beaker.

3. $(0.420 \text{ g} / 0.932 \text{ g}) \times 100\% = 45.1\%$
   The experimental result of 45.1% agrees with the theoretical value of 45.58%.

4. $(13.7 \text{ g} / 18.02 \text{ g/mol} = 0.760 \text{ mol H}_2\text{O})$; $(86.3 \text{ g} / 227 \text{ g/mol} = 0.380 \text{ mol AC})$

   After dividing mol H_2O by mol AC $(0.760 \text{ mol} / 0.380 \text{ mol})$, the water of crystallization is found to be 2, and the formula is $AC \cdot 2H_2O$.

5. The hydrate is completely decomposed when the moisture inside the beaker is gone. It will change in appearance from granulated sugar (*before heating*) to powdered sugar (*after heating*).

6. A warm beaker radiates heat and warms the air around the balance pan. This causes an error in weighing because warm air rises, thus lifting the balance pan.

7. • Not allowing the beaker to cool causes a light weighing, which gives *high results*.
   • Not removing water on the watchglass causes a heavy weighing, which gives *low results*.
   • Incomplete heating of the hydrate causes a heavy weighing, which gives *low results*.
   • Overheating the hydrate can decompose the anhydrous compound, which gives *high results*.

8. • Wear eye protection; heat the watchglass gently above a *low* burner flame to avoid breakage.
   • When weighing the beaker and watchglass, transport carefully so as to avoid breakage.

Experiment 14 — *Decomposing Baking Soda*

1. See the Glossary, Appendix I.

2. When all the baking soda is decomposed, carbon dioxide is no longer produced and the water level in the beaker remains constant. After the burner is shut off, the water level will actually decrease in the beaker as the gas is allowed to cool.

3 Yes, some errors can lead to high results, giving a percent yield greater than 100%.

4. $(0.955 \text{ g} / 0.9462 \text{ g}) \times 100\% = 101\%$

5. $(1.10 \text{ g} / 2.000 \text{ g}) \times 100\% = 55.0\%$

6. • Heating the baking soda mixture insufficiently leads to high results.
   • Weighing the test tube containing traces of moisture from the decomposition of baking soda gives a heavy mass reading.
   • Weighing the test tube while warm gives a weighing error due to a light mass reading.

7. • Wear eye protection; be careful when heating the substances in a test tube.
   • Avoid pinching off the rubber tubing leading from the test tube to the Florence flask.

Experiment 15 — *Precipitating Calcium Phosphate*

1. See the Glossary, Appendix I.

2. Particles of precipitate will be transferred into the filter paper and slow the rate of filtration.

3. To thoroughly remove the precipitate from the inside of the beaker.

4. If particles of precipitate appear in the filtrate, you may have to recycle the filtrate through a second disk of weighed filter paper (ask the Instructor). If two filtrations are required, find the mass of precipitate from each filtration and add the two masses of precipitate together.

5 Yes, some errors can lead to high results giving a percent yield greater than 100%.

6. $(0.466 \text{ g} / 0.475 \text{ g}) \times 100\% = 98.1\%$

7. $(0.542 \text{ g} / 0.995 \text{ g}) \times 100\% = 54.5\%$

8. • Incomplete precipitation gives low results.

   • Coprecipitation of impurities gives high results.

   • Particles of precipitate in the filtrate give low results.

   • Weighing filter paper with precipitate before it is completely dry gives high results.

   • Weighing warm filter paper before it cools gives low results.

9. Wear eye protection; be careful when handling glassware and using the laboratory burner. Dispose of the precipitate and filter paper as directed by the Instructor.

Experiment 16 — Generating Hydrogen Gas

1. See the Glossary, Appendix I.

2. If the magnesium has a mass greater than 0.09 g, it may produce a volume of hydrogen greater than the capacity of the 100-mL graduated cylinder.

3. Small gas bubbles are no longer released.

4. When hydrogen gas is collected over water, the hydrogen contains water vapor and is referred to as a "wet" gas. A "dry" gas does not contain water vapor. To obtain the partial pressure of "dry" hydrogen gas in this experiment, we must subtract the vapor pressure of water from the atmospheric pressure at the experimental temperature.

5. The partial pressure of the hydrogen gas is equal to the atmospheric pressure minus the vapor pressure of water at 23 °C. Table 16.1 lists the vapor pressure of water at 23 °C as 21 mm Hg. Therefore, the partial pressure of hydrogen gas is 756 mm Hg – 21 mm Hg = 735 mm Hg.

6. 80.7 mL H_2 at STP; 0.00364 mol H_2 at STP; 22.2 L/mol

7. The major sources of error in this experiment include:
   • an air bubble in the graduated cylinder after inverting the cylinder filled with water
   • incomplete reaction of the metal
   • misreading the meniscus in the graduated cylinder

8. • Wear eye protection; be careful with the glassware when collecting the gas.
   • Avoid contact with dilute hydrochloric acid.
   • Avoid breathing the fumes from hydrochloric acid.
   • Handle the thermometer carefully, as it is easily broken and can cause cuts.
     (Report a broken thermometer immediately to the Instructor; mercury vapor is hazardous.)

Experiment 17 — *Generating Oxygen Gas*

1. See the Glossary, Appendix I.

2. Manganese dioxide catalyzes a safe and rapid decomposition of potassium chlorate.

3. When all of the potassium chlorate is decomposed, oxygen is no longer produced and the water level in the beaker remains constant.

4. The experimental volume of oxygen gas produced is equal to the volume of water displaced into the beaker; for example, 205.0 g H_2O = 205.0 mL H_2O = 205.0 g O_2.

5. The partial pressure of the oxygen gas is equal to the atmospheric pressure minus the vapor pressure of water at 22 °C. Table 17.1 lists the vapor pressure of water at 22 °C as 20 mm Hg. Therefore, the partial pressure of oxygen gas is 765 mm Hg – 20 mm Hg = 745 mm Hg.

6. 252 mL O_2 at STP; 0.919 g $KClO_3$; 90.6%

7. • Heating the potassium chlorate mixture insufficiently can lead to high results.
   • Weighing the test tube while warm gives a weighing error due to a light mass reading.

8. • Wear eye protection; be careful when heating the substances in a test tube.
   • The potassium chlorate mixture must not contact the rubber stopper in the test tube.
   • Potassium chlorate must not be heated without manganese dioxide to moderate the reaction.
   • Avoid pinching off the rubber tubing leading from the test tube to the Florence flask.

Experiment 18 — *Molecular Models and Chemical Bonds*

1. See the Glossary, Appendix I.

2. The number of valence electrons correspond to the group number of the element; thus,

   (a) O = 6   (b) N = 5

   (c) Br = 7   (d) I = 7

3. (a) hydrogen atom   (b) carbon atom

   (c) oxygen atom   (d) nitrogen atom

   (e) single bond (2 e–)   (f) double bond (4 e–)

4. (a)   I—Br
   (b)

$$\begin{array}{c} H \\ | \\ H-C-Cl \\ | \\ H \end{array}$$

   (c)

$$\begin{array}{c} O \\ \| \\ Cl-C-Cl \end{array}$$

5. (a) $:\!\overset{..}{I}\!:\!\overset{..}{\underset{..}{Br}}\!:$
   (b)

$$\begin{array}{c} H \\ H:\overset{..}{C}:\overset{..}{\underset{..}{Cl}}: \\ H \end{array}$$

   (c)

$$\begin{array}{c} :\overset{..}{O}: \\ :\overset{..}{\underset{..}{Cl}}:\overset{..}{\underset{..}{C}}:\overset{..}{\underset{..}{Cl}}: \end{array}$$

6. (a) 14 e–; (b) 14 e–; (c) 24 e–

Experiment 19 — *Analysis of Saltwater*

1. See the Glossary, Appendix I.
2. The dark brown iodine crystal will impart a faint yellow color to the solvent.
3. Heating tap water leaves carbonate deposits on glassware.
4. The supersaturated solution should be stirred to provide a homogeneous solution. Disturbing the test tube, after it has cooled, may cause the solution to crystallize prematurely.
5. (a) 4.50%
   (b) 0.794 *M*
6. • Pipetting is the main source of error. A 10.0-mL sample of saltwater should weigh slightly more than 10.0 g, for example 10.214 g. Check your data, and pipet again if necessary.
   • An evaporating dish that has moisture on the bottom gives rise to high results.
   • Not allowing the evaporating dish to cool completely before weighing gives high results.
7. • Wear eye protection; be careful when using the laboratory burner, thermometer, and pipet.
   • Refer to the diagram of an evaporating dish on page 4, and avoid breakage during transport.
   • Keep organic liquids and vapors away from a burner flame.
   • Avoid breathing the vapor from the organic solvents.
   • Dispose of the organic liquids in a special waste container designated by the Instructor.

Experiment 20 — *Analysis of Vinegar*

1. See the Glossary, Appendix I.
2. (a) 0.50 mL   (b) 31.35 mL
3. As the endpoint is approached using phenolphthalein indicator, flashes of pink color persist longer. At the endpoint, it requires only 1 drop of NaOH (~0.05 mL) to "flip" the indicator, and change the titrated solution from colorless to permanent pink.
4. The volume of NaOH required to reach an endpoint will vary for each trial, depending on the mass of the KHP sample.
5. The volume of NaOH required to reach an endpoint should be about 30.15 mL for each trial because the amount of acetic acid is the same in each 10.0-mL vinegar sample.
6. 0.202 *M* NaOH
7. (a) 0.840 *M* $HC_2H_3O_2$; (b) 5.00% $HC_2H_3O_2$
8. (a), (b), (e)
9. • Wear eye protection; be careful when using the laboratory burner, pipet, and buret.
   • Add NaOH carefully to the buret through the funnel, and avoid overfilling with NaOH.
   • Avoid contact with NaOH. In the event of contact, wash the area immediately with water and notify the Instructor.

Experiment 21 — *Electrical Conductivity of Aqueous Solutions*

1. See the Glossary, Appendix I.

2. (g) is a gas; (l) is a pure liquid such as water; (s) is a solid or precipitate; and (aq) is an aqueous solution containing a dissolved substance

3. Refer to Table 21.1.

    (a) strong acids, strong bases, and soluble salts

    (b) weak acids, weak bases, and very slightly soluble salts

4. (a) A strong electrolyte produces a bright glow in the light bulb.

    (b) A weak electrolyte produces a dim glow in the light bulb.

5. (a) ionized $H^+(aq) + Br^-(aq)$; (b) nonionized $HF(aq)$; (c) ionized $Sr^{2+}(aq)$ and $2\ OH^-(aq)$;

    (d) nonionized $NH_4OH(aq)$; (e) ionized $Ag^+(aq)$ and $NO_3^-(aq)$; (f) nonionized $Ag_2SO_4(s)$

6. The electrodes and beakers must be rinsed with distilled water in order to avoid a false-positive strong conductivity test for a weak electrolyte.

7. • Wear eye protection; be careful to avoid contact with the chemical solutions.

    • Do not touch the exposed wire electrodes, as the wires can give a serious shock.

Experiment 22 — *Activity Series of Metals*

1. See the Glossary, Appendix I.

2. (a) 0   (b) –1   (c) +1   (d) +7   (e) +5

3. (a) $Cr_2O_7^{2-} + 6\ Fe^{2+} + 14\ H^+ \rightarrow 2\ Cr^{3+} + 6\ Fe^{3+} + 7\ H_2O$

    (b) $Mn^{2+} + H_2O_2 + 2\ OH^- \rightarrow MnO_2 + 2\ H_2O$

4. $Al > Cd > Ni > Ag$

5. • Wear eye protection; be careful to avoid contact with conc H_2SO_4 and conc HNO_3.
    (If contact occurs, wash immediately with water and notify the Instructor.)

    • Avoid breathing any of the following gases, which are highly irritating and require proper ventilation: SO_2, H_2S, NO, NO_2, and NH_3.

    • Dispose of potassium permanganate and other manganese-containing solutions in a special waste container designated by the Instructor.

Experiment 23 — *Organic Models and Functional Groups*

1. See the Glossary, Appendix I.

2. The elements represented by each colored ball are as follows: black—carbon, red—oxygen, green—chlorine, blue—nitrogen.

3. Two black balls joined by one connector represents a carbon–carbon single bond: C—C.

4. A black and red ball joined by two connectors depicts a carbon–oxygen double bond: C = O.

5. (a) methyl; (b) ethyl; (c) isopropyl; (d) propyl

6.

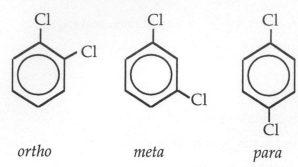

ortho	*meta*	*para*

7. (a) ether   (b) organic halide

   (c) amine   (d) ketone

   (e) carboxylic acid   (f) aldehyde

   (g) amide   (h) alcohol

   (i) phenol   (j) ester

Experiment 24 — *Separation of Food Colors and Amino Acids*

1. See the Glossary, Appendix I.

2. The "origin" is the line on a chromatogram where a drop of sample is spotted.

3. "Spotting" is the process of putting a small drop of sample on the chromatogram; usually a glass capillary tube is used to spot a chromatogram.

4. A "lane" is the imaginary path of the components in a sample as the solvent moves up the chromatogram. By convention, the lanes are numbered from left to right (e.g., 1 to 4); and component #1 in a given lane has moved the farthest and is closest to the solvent front.

5. The R_f value of the amino acid is: 2.1 ~~cm~~ / 7.0 ~~cm~~ = 0.30

6. A pen should not be used on a paper chromatogram because the ink in the pen can dissolve in the solvent and confuse the results.

7. • Wear eye protection; be careful to avoid cuts with the glass capillary tubes.

   • Do not use the organic solvent near a laboratory burner flame.

   • Avoid splashing the solvent when adding to the developing chamber.

   • To avoid a large initial spot, practice spotting on scrap chromatography paper.

   • Wear latex gloves when handling the amino acid samples and the chromatogram.

   • Avoid breathing and touching ninhydrin while spraying the chromatogram.

   • Dispose of the solvent in an organic waste container designated by the Instructor.